Projective Techniques and Sort-Based Research Methods

Projective Techniques and Sort-Based Research Methods offers a brief introductory guide to the use of these exciting, innovative and often artistic approaches to students and researchers who have no prior knowledge of these.

This book brings together a wide range of examples of projective and mapping techniques that offer the ideal methodology for researchers wishing to collect less controlled and filtered material that tap the deeper levels of the conscious and subconscious to reveal a more profound, richer and hidden level of response. It presents the techniques in a way that will enable the reader to appreciate their nature and to choose an appropriate method for their own research. Information is also provided that allows readers to design and implement their own projective or sort-based approaches. Each of the approaches the authors present is concisely described, and their usages are explained, along with references and examples of the applied usage of the respective technique.

The book is a valuable reading for researchers from a wide range of academic disciplines within the social sciences, humanities, business studies, marketing and so forth. The book is an introductory guide, but it will be appropriate for use with undergraduate, post-graduate and research students. It will also be of great use to professionals working in the areas of consumer behaviour, marketing and communications.

Paul M. W. Hackett is a professor and author with positions at universities in the US, the UK and Nigeria. He is the author of more than 250 publications, including over 25 books. He is the originator of the declarative mapping approach to social science and humanities research, with PhDs in both psychology and fine-art.

James M. Suvak is an Emerson College graduate. He majored in marketing communications and discovered his passion for research after taking a course with Paul Hackett. James now continues his journey in the field of research and marketing by working at a consulting firm in Boston.

Ava Gordley-Smith is a PhD candidate whose work focuses on theoretical and applied research methodologies in the social sciences and creative disciplines. She is engaged in several writing and research projects, utilising ecology-centric and sociological methodologies as an approach to constructing both marketing and brand strategies. Ava has over ten years of marketing management experience and an MA in strategic marketing communication from Emerson College.

Projective Techniques and Sort-Based Research Methods

Paul M. W. Hackett,
James M. Suvak
and Ava Gordley-Smith

Routledge
Taylor & Francis Group
LONDON AND NEW YORK

First published 2023
by Routledge
4 Park Square, Milton Park, Abingdon, Oxon OX14 4RN

and by Routledge
605 Third Avenue, New York, NY 10158

Routledge is an imprint of the Taylor & Francis Group, an informa business

British Library Cataloguing-in-Publication Data
A catalogue record for this book is available from the British Library

ISBN: 978-1-032-25967-3 (hbk)
ISBN: 978-1-032-25968-0 (pbk)
ISBN: 978-1-003-28589-2 (ebk)

DOI: 10.4324/9781003285892

Typeset in Times New Roman
by Apex CoVantage, LLC

Paul dedicates this book to Jessica.
James dedicates this book to Paul Hackett and Ava Gordley-Smith.
Ava dedicates this book to Emerson College, Paul Hackett and James Suvak.

Contents

Figures

Preface

Researchers who are interested in human behaviours and the ways in which human beings experience and make sense of their worlds are always faced with questions regarding the best or most applicable way to get hold of the information they want. The first question a researcher has to satisfy is whether the information they need to answer their queries already exists. If the required information does exist, then they may use this in their project. For example, if I was interested in the number of people who have died from COVID-19-related illnesses in 2021, this information is likely to exist, and if my research question was concerned with the differences in mortality rates between the UK and the US, then again, I could probably find valid and reliable data to answer such a comparative question.

However, if I was concerned not with international COVID-19 mortality rates but with how residents in a specific geographical area experienced and were personally impacted by having COVID-19, then this information may not be already available. Consequently, I would have to design a research study that would gather the information I needed to answer this question, first-hand from selected individuals who have had COVID-19 in the geographical region in which I was interested. As this therefore requires me to gather new or original data, the next question that I need to ask is how I should design my research project so that it will avail me of the type of information I need to answer my specific questions regarding these personal experiences. The pre-existing procedures that I employ, or novel approaches that I could develop, to gather the information that I desire will access responses and various aspects of an individual's experiences as assessed in a wide variety of ways.[1]

There are very many research approaches that may be used in social science and humanities research. A considerably large number of books also exist on social science research approaches in general and also on specific aspects and specialisations within this domain. However, when we did a comprehensive search of the literature, we discovered that there are practically no books that specifically and exclusively address the social science research approaches that fall under the banner of the particular form of research method known as projective techniques. There are chapters and sections of larger textbooks that do concern themselves with projective methods, but this is scant coverage of the necessity of these being small sections in larger texts. As well as these chapters and a number of journal articles, there are also a small number of rather old and obscure books on the topic. It is therefore the authors' main objective to fill this omission in the literature.

As well as filling this gap in the publications that exist on projective techniques, we, as authors, had several objectives and desires for the text they set themselves when they started to write this book. Overall, it was our aim to write a book that was easily accessible to nascent researchers and also to those with only minimal research experience. That is to say, we wanted our text to be easily accessible to a wide range of students, academics and professionals without the reader having to supplement our book with reading about research methods using other books or articles. We also wanted our book to form a *go-to* text: a single volume to which any person wishing to use a projective approach in their research design would turn. With these aims in mind, we, therefore, decided that the idea of providing a comprehensive review of all, or a large number of, projective techniques would be out of the question as such a book would be large, unwieldy and not easy to use. This size of such definitive and comprehensive text may also put novice researchers off attempting to even take the book off the shelf! As a consequence of these initial aims and stipulations for the text, early on in the writing of this book, we decided to be extremely selective in the examples of the different projective approaches that we included in the book. Selecting projective approaches was not a simple task as these are so diverse in their content and the ways in which they are used.

It is our belief that projective techniques are particularly useful forms of social science research methods for several reasons. The first of our justifications for viewing projective techniques in such a positive manner comes from the fact that the use of projectives in research grew out of their use in psychotherapy. In these forms of therapeutic situations, projective techniques are used during client and patient interactions in order to uncover or reveal latent, suppressed and other content of which the client may not be consciously aware. The term "projective" comes from this therapeutic context and refers to the notion that if a client is presented with a stimulus that is to them neutral (neutral in reference to the issues the client is seeking help with), then they will project repressed emotions and thoughts upon these stimuli. Another reason for our advocacy of projective techniques is due to our belief that these procedures have the potential of minimising the guiding role of the researcher and putting the respondent firmly in control of their own responses. In the following section, we will outline the contents of our book.

Paul M. W. Hackett

Note

1 In attempting to access such information, we may employ procedures that delve into short-term, long-term and working memory, an individual's visual abilities/spatial abilities and verbal (often word-based) abilities.

Acknowledgements

Paul's original idea for a book about projective techniques grew out of an eventual realisation that there was a need for this book. The recognition of this requirement has developed through his interactions with the hundreds of students that have been in his research classes and who have illustrated to him the need for a book that is dedicated to projective techniques. It has been these students' enquiring minds that engaged with the process of research, along with his use of projective techniques in his own research, that convinced him of the value of such a participant-driven form of research method. Paul is exceptionally grateful to his two colleagues/co-authors, James M. Suvak and Ava Gordley-Smith, who formed the team that transformed this book from an idea into a reality.

Ava M. Suvak acknowledges her deep appreciation for the support, effort and academic rigour of Paul M. W. Hackett and James M. Suvak during the researching and writing of this book. She also wants to acknowledge Emerson College, the influential institution that bound them together by the common thread of their passion for projective techniques.

Ava Gordley-Smith acknowledges Emerson College as this book would not have been possible without it. Additionally, James acknowledges the immense support of Paul M. W. Hackett and Ava Gordley-Smith throughout the writing process.

Introduction

In the first chapter "What Are Projective Techniques, Mind Mapping and Sort-Based Methods?", we present an introductory outline of projective techniques and their roles in social science and humanities research. Furthermore, in this introductory chapter, we give a brief history of projective approaches and provide a theoretical background to these research methods, which involve us noting the roots of these approaches in psychotherapy. It is our intention in this first chapter to provide enough details to encourage readers with a little background in this area to continue reading further into the book. We also hope our writing will be at a level that engages the more research-experienced reader. In the second chapter, we go into projective details in greater depth and note the major classes of projective research techniques such as mapping procedures and sort-based techniques. In the third chapter, we concentrate upon one of the main forms of projective techniques: sort-based procedures. We provide further details of how these approaches are used, the types of information they produce and the sort of questions they are most suitable to be used with. In chapter four, we address another major class of projective research techniques: mapping procedures. In chapter five, we turn our attention to the nitty gritty of how a researcher can develop their own projective technique or approach. This chapter provides detailed information for researchers of all levels who desire to use a projective means of data gathering but cannot identify a pre-existing procedure that they feel exactly

DOI: 10.4324/9781003285892-1

fits their research's needs. In the sixth chapter, we attempt to see into the future with our crystal balls in the hope of predicting the direction in which we believe projective approaches to research may progress in the future, and we make suggestions in regard to the research areas in which we believe projective approaches may be beneficially used. We start by suggesting that projectives are especially suited for use in critical social research, which attempts to uncover systemic and individual biases against people based upon race and other characteristics. We also suggest that, due to the ways in which projective techniques are particularly responsive and sensitive to input from participants, projective techniques have a potential utility and are especially suited when conducting research into sensitive research topics and with individuals from marginalised communities. We next consider using projectives within the conceptual system and framework for conducting social science research known as the declarative mapping approach, and finally, we consider how since the COVID-19 pandemic, much human subject research has moved online and suggest the potential usefulness of projective approaches as digital research methods. In chapter 7, our final chapter, we provide a terse synopsis of the content of the book's content. Whilst chapter 7 is the final chapter, we have two sections that follow it. First section provides a short reading list of books and articles that we believe would be of both interest and benefit to readers who wish to delve a little more deeply into projective approaches. Finally, we offer a glossary of the important research terms in this book and other important research terms. We provide the glossary because, as we stated in the intentions for the book in this preface, we wish that this book stand alone as a guide for using projective techniques for novices to these approaches and indeed novices to social science research in a general sense. The glossary is therefore included to enable neophyte researchers to clarify terms they may not be familiar with without having to look these up externally to our text. We have also included some terms that do not appear in the text but that we feel are important terms and concepts that someone using a projective technique may encounter.

It should be noted that this book has been written with a large and broad audience in mind. The book should appeal to many academic disciplines within the social sciences and humanities. The book will also be of use and interest to professionals in marketing research and marketing fields.

The authors of this book have different degrees of contact and types of experiences with projective techniques. They also have different academic and professional backgrounds and experiences. However, one thing we authors have in common is that we are passionate about conducting social science and humanities research and through our encounters with projective research have developed a respect for these techniques.

1 What Are Projective Techniques, Mind Mapping and Sort-Based Methods?

Chapter Summary

In this chapter, it is appropriate that we commence by introducing readers to the subject matter of the book, namely, projective techniques that may be used in social science and humanities research and by professionals in marketing. It also seems appropriate that we commence this introduction by considering where projective techniques come from, their history and theoretical background, which will include a brief reflection upon their origins in psychoanalytic theory, psychoanalysis and therapeutic approaches. We hope that this book will be used as a guide to using projective approaches by those who have little or no background in using projectives. Therefore, this chapter aims to whet the appetite of those coming newly to projectives, and we hope that rewarding this will make them want to learn more about these and to read further into this book. It is also our intention that there we will provide enough detail so as to engage the researcher who is somewhat experienced with projective techniques, and that they will also find the book interesting and of use to them.

Introduction

This is an exciting and fascinating book!

This is a bold claim that we make with conviction. The reason for our confidence is that the subject matter of this slim

DOI: 10.4324/9781003285892-2

volume is something with which we are all concerned and is something that human beings have spent centuries trying to understand. Indeed, we have even conjured up gods, demons and other embellished and exaggerated, mysterious and wonderful processes to provide us with an understanding of what may be called hidden motives or why we do what we do and make the choices we make. We have a desire to predict and understand human behaviour, and we may ask for instance:

- Why did you vote for Biden rather than Trump or Johnson rather than Corbin, Truss rather than Sunak?
- Why do you drive the make and model of that car you drive?
- Why do you think the Thingvellir valley is the most beautiful place you have ever visited?
- How do you choose between going out to the movies or staying in tonight?

All of the questions and the millions and millions of other questions that human beings resolve every day are, to a greater or lesser extent, asked and answered subconsciously. We are constantly attempting to predict the world around us by gathering information through our senses and relating this to our memories so that we may foretell what will happen in the future. So, it does seem reasonable to say that this book, concerned as it is with attempting to reveal the human motivations that underlie the decisions that we make on a daily basis, addresses issues that are of great interest to us all.

When we think about it, there are many ways through which we may attempt to discover why someone else does what they do. The most direct approach that is available to us is simply to ask them. However, this assumes that they know why they did the thing we are interested in discovering their motives about, and that they will report what they know honestly to us. Another way that we may attempt to understand human behaviour is to accurately observe someone's actions. However, simply seeing someone do something may allow us to understand what they did but tells us little as to why they did this and if they are likely to do this again. This then puts us in a position where we may have to follow up our observations by asking a person why they acted as they did. Furthermore, whether we directly ask someone a question, in person or by using a questionnaire, if we observe someone committing an action or if we employ a myriad of other data-gathering techniques, it is implicitly assumed that the person we are conducting research upon is consciously aware of their motivations and that they are willing and able to accurately report these to us. In reality, this assumption is often not upheld. The reason for this inaccuracy is down to many reasons. For example, someone may buy a certain make and model of car, and when asked why they purchased this specific car, they may cite performance figures, depreciation costs being low, safety features and so forth. However, they may well have purchased this car simply because they liked the colour or wanted to impress their neighbours. The inaccurate statements that someone may make can be for several reasons. For instance, a person may make a report of their motives in order to impress us as researchers, to appear politically correct, environmentally aware, wealthy, for fun and so on. Inaccuracy in the reports we may receive may also stem from their subconscious responses and the individual we are talking to may be positively or negatively inclined towards a specific response and may not even be aware of this inclination: for example, they may just have an almost visceral reaction to the car.

As a consequence of these and other difficulties associated with the validity of verbal reports given to by respondents, researchers decided that it would be beneficial to develop techniques and methods that could avoid these inaccuracies. One of the attempts that researchers have made in this regard is the creation of methods that aim to tap into the sub- or pre-conscious reactions and motivations that individuals have to the world around them. These forms of data-gathering techniques are the subject of our writing in this book and are known as projective techniques, mind mapping techniques and sort-based methods.[1]

Within anthropology, it is often said that the aim of the research is to look at our everyday activities and try to make the familiar strange and the strange familiar (Miner, 1956), and whilst this statement is not directly referring to projective research techniques, the aim behind the use of projective approaches is similar. The techniques we present in our book also attempt to adopt what anthropologists call an emic perspective rather than an etic approach, where the emic types of research attempt to present the perspectives of those being studied. Conversely, an etic approach offers the perspective of those making the observations or doing the research (e.g. see Dundes, 1962; Jingfeng, 2013). Projective techniques, mind mapping and sort approaches all make efforts to offer a deeply personal and emic perspective of those participating in the research.

Synopsis of Contents

In our writing in this first chapter, we will start with an introduction to projective techniques which will be followed by a brief presentation of the history of projective techniques. In the next part of the chapter, we review the origins in psychoanalysis and therapeutic approaches of the techniques we consider. Next, we look at the background of projective techniques, including their theoretical bases. Finally, we sum up the chapter with some concluding thoughts.

This book is intended as a guide to using the aforementioned techniques. We hope that readers who have no experience at all with these approaches to research will find the book interesting and stimulating and impel them to use these data gathering in their own research. It is similarly our ambition that more experienced researchers will also find the book interesting and of use to them. As our writing is primarily intended to first introduce and then provide usage suggestions for projective and other techniques to those who are interested in using these, the text will concentrate on application. The theory and history that underpin the approaches will be provided in such a way that these will assist readers to understand the uses and limitations of the approaches and will enable them to better contextualise the approaches we cover.

Within this book, we conceive of projective techniques in a broad manner, and in this first chapter, we will provide a background and an introduction to the projective techniques that we will include in our considerations. This introduction will take the form of a presentation of the background and theory to the many projective techniques, sort procedures and mental mapping as data-gathering approaches that exist. A very brief history of projective techniques will be provided as will their origins in psychoanalytic theory, psychoanalysis and as therapeutic approaches, and we will touch upon and present some of their academic psychological origins in works such as George Kelly's Personal Construct Theory (Bannister and Fransella, 2019; Kelly, 1963, 1992a, 1992b, 2019; Winter and Reed, 2015) and the repertory grid technique (Fransella, 2003; Hayes, 2013).

Introduction to Projective Techniques

The projective techniques methodology is a form of qualitative scientific research (Nunez, 2018; Projective Methods) that aims at uncovering the unconscious motivation and decision-making process. These approaches are designed to stimulate a respondent through the presentation to them of potentially ambiguous stimuli in order to discover deep-rooted and concealed proclivities, behaviours and various other forms of personal data. Psychology research, and indeed our own personal experiences, has led us to believe that human beings lack true rationality and quite often have a limited awareness of their own subconscious. This insufficient self-awareness may often result in the inability for the individual to grasp their own reasonings and desires. Projective techniques assist to bring forth these obscured motivations and values and may enable us to gain a more comprehensive understanding of a behaviour as a whole. Projective techniques are most widely used in psychoanalytic theory, clinical psychology, social psychology and cultural anthropology.

Although there are many tactical differences amongst the techniques themselves, the commonalities shared are sufficient enough to stand them apart from alternative personality measures, such as inventories, ratings or situational tests (e.g. see: Chamorro-Premuzic, 2015; Butcher, 2010). Projective techniques collectively share a relatively unstructured and ambiguous nature. These shared frameworks ultimately require the participants to apply meaning, therefore revealing fundamental information about themselves.

Historically, Projective techniques have been categorised into five groupings: association techniques, construction techniques, choice or ordering techniques, expression techniques and completion techniques (Linzey, 1959). These techniques may be exemplified in the Rorschach's Inkblot Test (Rorschach, 1921), Thematic Apperception Test (TAT) (Aronow et al., 2001; Reznikoff, 2001), Draw-A-Person Test and Sentence Completion Test. The following section of this book will aim to offer a brief overview of the historical markings in the development of Projective techniques.

History of Projective Techniques

The genesis of projective techniques arguably dates back to the creativity of the Neanderthal species of humans in cave paintings at least 64,000 years ago. People's fascination with drawing deeper meaning from visual stimuli such as art, clouds and dreams (Freud, 1953) reveals a desire to gain a deeper understanding of themselves, and reflexively reveals the ability for ambiguous stimuli to unearth these greater contexts (Klopfer, 1973). The contemporary usage of projective techniques began to prosper in the latter half of the 19th century, woven closely within the use of psychoanalysis (Rabin, 2001). Psychologist Sir Francis Galton developed the word association method as a way to test and measure intellect. Galton (1879) gave the respondent a stimulus word and four seconds to determine as many associated words as they could. Although he failed to connect intellect to this particular analysis, his work sparked the continued development of projective techniques as a form of measurement (Simonton, 2003). Inspired by Galton's work, the prominent Swiss psychiatrist and psychoanalyst Carl Jung helped to standardise the word association technique as a form of projective testing. Jung's test consisted of 1,000 stimulus words and asked participants to answer with the first words that came to mind as fast as possible. According to Jung (1910), those who responded with significant delay would be likely to have a "complex" moreover an "inferiority complex" from an emotional disturbance. Circa the same time Freud was theorising the analysis of dreams and developing his dominant psychoanalytic method of dream interpretation (Rorschach, 1921).

The turn of the century marked a large leap forward for projective techniques. The period following World War I instilled momentum in the development of aptitude and ability tests as exemplified by the Stanford-Binet and various group tests. The clinical psychology usage of projective techniques dates back to the early 20th century and was first noted in 1925 by Hermann Rorschach's book *Psychodiagnostik*, detailing his famed Inkblot Projective Test technique. Rorschach was a follower of Carl Jung and agreed with Jung's psychoanalytic practices and opinions. Like Jung, Rorschach believed that the unconsciousness held people's innermost conflicts and moreover that these might be unveiled through one's reaction to ambiguous stimuli. Rorschach's Inkblot Test (see Figure 1.1) aimed to predict the nature of one's personality such as a measure of introversion or extroversion by observing the respondent's response to an image of a blot of ink on paper. Rorschach's work led to numerous adaptations of his methods.

Not long after Rorschach's developments, Henry Murray and Christiana Morgan (Douglas, 1993) developed the TAT in 1935. Unlike Rorschach's Inkblot Test which sought to analyse and reveal the uncommon facets of the participant's personalities, the TAT looked to measure the common traits. The TAT presented images to the respondent illustrating one or more individuals interacting ambiguously and then asked the participant to develop a story for the individual(s) in the

Figure 1.1 Example of Rorschach's Inkblot Test

image. The stories told were to identify needs such as achievement and companionship (Klopfer, 1973).

Sentence completion became more prominent in the first few decades of the 20th century. An example of this might look like, "I am sad because _____. This technique was essentially offering the stem of a sentence to the participant and then asking them to complete it. An example of this might look like, "I am happy with _____" (Rhode, 1957). This method intends to reveal underlying fantasies, thoughts or emotional conflicts more than would happen through the presentation of a direct question (Weiner and Greene, 2008). Carl Jung is noted as the first to use sentence completion techniques to analyse respondents' personalities (Hersen, 2003). Although widely used, in the 1950s, this form of technique became controversial for its inconsistent and difficult ability to gather standardised measurements. This growing concern sparked an increased demand for standardisation and continued iterations.

Goodenough and Harris developed the Goodenough-Harris Test in the 1930s as an approach to measure intelligence through the expression of drawing (Meyer et al., 2020). Buck amended this form and introduced his version as the House-Tree-Person Test in 1949. This test consisted of asking the participants to draw a house, a tree and a person and would aim to once again reveal intelligence. Karen Machover also modified the Goodenough-Harris Test and created the Draw-A-Person Test, which aimed to analyse the personality of the respondent by asking them to draw a man, a woman and themselves in the picture. Machover (1949) additionally published a book to assist this process. See Figure 1.2 for examples.

Figure 1.2 Examples of House-Tree-Person Test

Further Background to Projective Techniques

Projectives' origins lie within the field of psychology and are commonly used in qualitative research, especially as a part of a focus group or an in-depth interview. The goal in using projective tests is to extract deep-seated attitudes and motivations that are often not revealed through traditional questioning. Projectives (also referred to as projective techniques or projective exercises) typically encourage respondents to open up and express their emotions by participating in a creative activity or inventive game. Imaginatively designed projective exercises that place respondents in amusing situations and environments tend to motivate them to drop defence mechanisms and disclose raw feelings and thoughts about the topic in question. Projective techniques allow researchers enter the subconscious mind of respondents by breaking down their fortifications, which empowers respondents to divulge their subjective or true opinions and beliefs about brands, events, products, services, concepts, objects and so forth.

Projective exercises are not limited to being used when direct inquiries are met with hesitant answers from respondents. For instance, respondents may be sometimes unconscious in regard to their beliefs or motivations, could have trouble articulating these or might be cautious when answering direct questions due to fear of societal norms. To combat these issues, some projectives are composed using a third-person approach in order to make respondents feel less confronted with the line of questioning associated with a research exercise. Hopefully, by not addressing direct questions to participants, they may, almost inadvertently, reveal their true opinions and feelings.

Projective techniques can be used in countless types of studies and applications. However, they tend to be most advantageous in situations that require the unveiling of subconscious behaviours, perceptions and values. Projectives can be beneficial in qualitative research settings, and researchers should give serious thought to the possible benefits of their using one or many exercise(s) in either a focus group, an in-depth interview or as a stand-alone approach to gathering information. When it comes to implementing projective techniques into a study, researchers should carefully examine and select between the different projective approaches that are available in order to end up using one that is designed to reveal information appropriate to their enquiries. Determining the appropriate exercise for the session to achieve quality results is difficult, but a researcher's understanding and analysis of the different forms of projectives available will improve through time and experience.

When researchers share with participants their intent to use a projective exercise in a session, the procedure may be explained in a manner that clearly describes the procedure to the participant whilst not being overly prescriptive in terms of what they expect the participant to produce. This will then allow ambiguous stimuli to be presented and for the participant, whilst understanding the procedure they are completing, to be placed into a situation that lets them "project" their thoughts and feelings onto neutral stimuli. The outcome and answers from these exercises are commonly followed up on with a more in-depth dialogue and discussion between the researcher and the participant. Generally, when projectives are inserted into a session, respondents are instructed to work individually. This is a crucial step in the process, as it ensures that participants will not only offer their own unique interpretations of the specific projective exercise but also not influence other participants' responses and any bias will be eliminated.

Projective Techniques and Psychological Theory

The theoretical assumption that underlies projective techniques is that when a particular question is proposed, the answer will be carefully and consciously constructed and socially determined. These responses may therefore not accurately represent the participant's unconscious or hidden motivations or attitudes. The respondent's true opinions and beliefs may not be cognisant of the respondent or the respondent may not be able to precisely articulate them in a satisfactory manner to the questioner. Projective techniques consist of using ambiguous stimuli or situations in order for the respondents to "project" their personality, opinions, attitudes or beliefs to shape the situation (Donoghue, 2000). It is believed that projective exercises may help uncover participants' concealed or latent thoughts and feelings about a specific situation or stimuli. These feelings and thoughts are unique to the respondent: in other words, the participants express their individuality. In addition, projectives are used to unearth respondents' own perspective on their world and how to act and behave in it. The concept of "projection" is commonly viewed as a defence mechanism that people use to protect themselves by compartmentalising their more cavernous thoughts and feelings (Gordon and Langmaid, 1988, p. 95). The theory of projective techniques is that they are based on the idea that unconscious beliefs and behaviours can be inferred by presenting a participant with ambiguous stimuli to encourage them to deploy their defence mechanism of projection. The respondent has the freedom to interpret and answer the stimuli (projection technique) in their own way (Loudon and Della Bitta, 1993, p. 300). Since there are no right or wrong answers, it is expected that participants will project their unconscious thoughts and feelings in their responses.

Having briefly described the background and rationale to projective techniques, we will now consider the background and rationale to two of the major sub-types of projective techniques, namely, sort procedures and mapping (Lemov, 2011).

Background of Sort Procedures

Sorting, by definition, is the process of systematically arranging, grouping and categorising items according to particular characteristics. There are many forms of sorting procedures that are used in research but the most well-known is probably the procedure designed by William Stephenson in the late 1930s, Q-methodology. Q-methodology, also referred to as Q-sort, was originally created with the intention of investigating "subjectivity" or perspectives, attitudes, viewpoints and motivations. Q-methodology is a research method that originates from psychology and social sciences and was developed in order to be used in both clinical settings and research settings. This methodology has proved to be valuable in a qualitative research setting as it allows researchers to understand respondents' processes of understanding and thought. The ultimate goal of these techniques is to aid researchers' analysis of how people evaluate and categorise pictures. In broader terms, Q-methodology provides a unique way of using visual stimuli to examine participants' mental concepts. Overall, sort methodologies such as card-sorting or picture-sorting do not require verbal responses but instead, visuals or photographs are given to respondents, who will then sort them into categories and groups, and rank them in order or identify any commonalities between them, depending on the sorting method in use. We have spoken in the above paragraph, but all forms of sort procedure are essentially similar in their overall or fundamental approach.

Background Mapping Techniques

Traditionally, methodologies used in qualitative research to collect and analyse data consist of observational methods, focus groups, in-depth interviews and many other methodologies. Interviews are one of the most common approaches to collecting data in qualitative research sessions. By definition, an interview is a verbal exchange of information between one or more respondents and the researcher(s). While there are numerous ways to conduct an interview (over the phone, in-person, or virtual), there is a general consensus about how the data collection is conducted. However, this is not the case when it comes to the data collection using the creation of maps in qualitative research. The development of mind maps by respondents who were flexible and versatile was of gathering information that are rooted in the idea of respondents creating visual maps of the research subject areas. Mind maps are best understood as diagrams which are employed to represent a research domain through the use of words, concepts and ideas arranged spatially around a central theme, word or concept. Mind maps are based upon the notion that participants are able to spatially express their understanding of the similarities and differences of the sub-aspects of a research topic in a performative way that reveals their attitudes, behaviours and general understanding of the research topic.

Mental maps stem from behavioural geography. By definition, mental mapping commonly referred to as cognitive mapping is the production of several psychological processes that take in, record, store, analyse and organise all information that a person observes about their spatial environment in their everyday life. For instance, when a researcher decides to deploy mental maps as a technique in their study, they are actually attempting to map the maps we have created in our minds throughout the day. The goal here is being able to successfully map these individuals' mind maps to be interpreted with the hope of yielding powerful insights relevant to the research topic.

In our minds, our frequent use of a physical area defines and redefines the picture we have of our environments. For example, let's look at your commute to work versus a weekend trip from Boston to Cape Cod. You commute to your office on State Street in Boston every morning by train and it takes you roughly one hour to get there. That being said, it is a destination that is visited on a recurring basis and it may be perceived as a shorter distance. On the other hand, the weekend trip to Cape Cod is quite a rare experience. This may result in the individual perceiving the distance of the trip to be a lot farther compared to their commute to work, even if the total travel time is only thirty minutes longer. We perceive our personal environment as more favourable and comfortable than other areas that are unknown, or to which we do not frequently travel, to be peculiar, strange and even risky.

Conclusion

In this introduction, we have set the stage for our later discussion about projective techniques and we hope that we have you at the edge of your seat in anticipation. In this chapter, we introduced you to a brief history of projective techniques and traced the roots of these approaches within psychoanalysis and psychotherapy. We attempted to acclimate you with the general concepts surrounding projective approaches and uses in such a manner that you now feel familiarised and engaged with projectives and are eager to learn more. Whilst this chapter worked to establish a base of knowledge, the following chapters will delve much more deeply into the details and applications of these fascinating research methods. In the next chapter, we will start to unpack the different projective approaches that may be used to gather information about human behaviour and experience.

Note

1 Projective techniques may be given on their own, in a focus group or interview, or they may form a part of a test battery.

References

Aronow, E., Altman-Weiss, K., and Reznikoff, M. (2001). *A Practical Guide to the Thematic Apperception Test: The TAT in Clinical Practice*. Philadelphia, PA: Taylor & Francis.

Bannister, D., and Fransella, F. (2019). *Inquiring Man: The Psychology of Personal Constructs* (3rd edition) (Psychology Library Editions: Personality). London: Routledge.

Butcher, J.N. (2010). Personality Assessment From the Nineteenth to the Early Twenty-First Century: Past Achievements and Contemporary Challenges. *The Annual Review of Clinical Psychology*, 6, 1–20. https://doi.org/10.1146annurev.clinpsy.121208.131420.

Chamorro-Premuzic, T. (2015). *Personality and Individual Differences*. London: John Wiley and Son.

Donoghue, S. (2000). Projective techniques in consumer research. *Journal of Ecology and Consumer Sciences/Tydskrif vir Gesinsekologie en Verbruikerswetenskappe*, 28.

Douglas, C. (1993). *Translate This Darkness: The Life of Christiana Morgan*. New York: Simon & Schuster.

Dundes, A. (1962). From Etic to Emic Units in the Structural Study of Folktales. *Journal of American Folklore*, 75(296), 95–105. https://doi.org/10.2307/538171, JSTOR i223629.

Fransella, F. (2003). *A Manual for Repertory Grid Technique*. New York: Wiley.

Freud, S. (1953). *The Interpretation of Dreams*. London: Hogarth Press.

Galton, F., F.R.S., (1879). Psychometric Experiements, *Brain*, Volume 2, Issue 2, London, England: William Clowes and Sons.

Gordon, W., Langmaid, R., and Mills, C. (1988). *Qualitative market research: A practitioner's and buyer's guide* (pp. 20–32). Aldershot, UK: Gower.

Hayes, N. (ed.) (2013). *Doing Qualitative Analysis in Psychology*. London: Psychology Press.

Hersen, M. (2003). *Comprehensive handbook of psychological assessment volume 2: personality assessment*. Hoboken, NJ: John Wiley and Sons.

Jingfeng, X. (2013). *An Anthropological Emic-Etic Perspective on Open Access Practices*. Bingley, UK: Academic Search Premier.

Jung C.G. (1910). The association method: lecture III. *The American Journal of Psychology*, 21(2).

Kelly, G.A. (1963). *A Theory of Personality: The Psychology of Personal Constructs*. New York: W. W. Norton & Company.

Kelly, G.A. (1992a). *The Psychology of Personal Constructs: Volume One: Theory and Personality*. London: Routledge.

Kelly, G.A. (1992b). *The Psychology of Personal Constructs: Volume Two: Clinical Diagnosis and Psychotherapy*. London: Routledge.

Kelly, G.A. (2019). *Inquiring Man: The Psychology of Personal Constructs* (3rd edition) (Psychology Library Editions: Personality). London: Routledge.

Klopfer, W.G. (1973, January). The Short History of Projective Techniques. *Journal of the History of the Behavioral Sciences*, 9, 60–65. https://doi.org/10.1002/1520-6696(197301)9:1<60::aid-jhbs2300090107>3.0.co;2-h. PMID: 11609454.

Lemov, R. (2011). X-rays of Inner Worlds: The Mid-Twentieth-Century American Projective Test Movement. *Journal of the*

History of the Behavioral Sciences, 47(3), 251–278. Hoboken, NJ: Wiley Periodicals, Inc.

Linzey, G. (1959). On the classification of projective techniques. *Psychological Bulletin*, 56(2), 158–168.

Machover, K. (1949). *Personality Projection: In the Drawing of a Human Figure*. Springfield, IL: Charles C Thomas Publisher.

Meyer, G. et al. (2020). The Effect of CS Administration or an R-Optimized Alternative on Potential Projective Material in Rorschach Responses From Six Studies and a Meta-Analysis of Their Findings. *Journal of Personality Assessment*, 102(1), 135–146. https://doi.org/10.1080/00223891.2018.1492926

Miner, H. (1956). Body Ritual Among the Nacirema. *American Anthropologist*, 58, 503–507. https://doi.org/10.1525/aa.1956.58.3.02a00080

Nunez, K. (2018, January 24). *Projective Techniques in Qualitative Market Research*. AMA San Diego. Retrieved from https://sdama.org/knowledge/projective-techniques-qualitative-market-research/

Projective Methods. *International Encyclopedia of the Social Sciences*. Retrieved January 24, 2022 from Encyclopedia.com: www.encyclopedia.com/social-sciences/applied-and-social-sciences-magazines/projective-methods

Rabin, A. (2001). Projective Techniques at Midcentury: A Retrospective Review of An Introduction to Projective Techniques by Harold H. Anderson and Gladys L. Anderson. *Journal of Personality Assessment*, 76(2), 353–367. https://doi.org/10.1207/S15327752JPA7602_15

Reznikoff, K.M. (2001). *A Practical Guide to the Thematic Apperception Test: The TAT in Clinical Practice*. Philadelphia, PA: Taylor & Francis.

Rhode, A. (1957). *The sentence completion method: it's diagnostic and clinical application to mental disorders*. New York, NY: The Ronald Press Company.

Rorschach, H. (1921). *Psychodiagnostik Talfen*. Bern: Huber.

Simonton, D. K. (2003). Francis Galton's *Hereditary Genius*: Its place in the history and psychology of science. In R. J. Sternberg (Ed.), *The anatomy of impact: What makes the great works of psychology great* (pp. 3–18). American Psychological Association.

Weiner, I. B., & Greene, R. L. (2008). *Handbook of personality assessment*. New York, NY: Wiley.

Winter, D.A, and Reed, N. (2015). The Wiley Handbook of Personal Construct Psychology, London: John Wiley and Son.

2 Projective Techniques

Chapter Summary

Having set the scene in chapter one for the rest of the book and having provided the reader with a general understanding of the projective approach, in this chapter, we turn to briefly reviewing a series of different kinds of projective techniques that may be used in research projects. The different types of approaches we present are text-based projectives, verbally based projectives, visually based projectives, mapping techniques and sort methodologies. We discuss the benefit these approaches have for unencumbered knowledge development within qualitative research and begin to address the notions of how these might be integrated into research design.

Introduction

Congratulations, you have made it to chapter two and have officially opened your eyes, perhaps for the first time, to the world of projective techniques. This chapter illustrates the various approaches that fall under the banner of projective techniques and will equip you with a general understanding of what they are and how to use them. It must be noted, however, that projective techniques are not fixed entities. There is not an absolute system for classifying these techniques, and the different types of projective techniques may often seem to simultaneously encompass different approaches to gaining

DOI: 10.4324/9781003285892-3

insight into the person being studied. Furthermore, projective techniques continually grow and are adapted to new situations, new research questions and new ways of presenting the technique to an individual and encouraging their interaction with the technique. Perhaps more than any other form of data gathering, projective techniques may be constructed specifically for a specific research project. However, with these points in mind, commonalities amongst projective techniques are readily discernible, and throughout this book, we divide projective approaches into the following sections:

- Text-Based Projectives: approaches that use writing or texts to elicit information from participants
- Verbally Based Projectives: the use of spoken approaches to facilitate data gathering
- Visually Based Projectives: approaches that employ visual stimuli, such as pictures, or physical items, as a basis for gathering data
- Mapping Techniques: asking participants to produce maps or spatial arrangements of concepts in a map-like configuration
- Sort Methodologies: approaches that ask respondents to sort things or concepts into groups or spatial arrangements
- Other Projective Approaches: approaches that do not fall under the above categories

The Importance of Context

Before we progress with our consideration of projective techniques under the above headings, we want to make a point and ask you to reflect upon this. The point is that traditionally, a dichotomy has been established between the researcher and the person who is the participant, subject or in some other way providing information for the researcher. In all forms of social science research and research involving human subjects, we would argue that such polar differences between the person doing the research and those providing responses are at least tenuous if not fallacious. All research is conducted within a physical, temporal and cultural context. Research that is conducted between a researcher and a participant also involves a social context. Moreover, the information that is gathered from participants is subjective and richly interlaced with the experiences of the respondent. The information that is gathered is also understood within a wide variety of contexts. This is the case with social science research when broadly understood, but many of these subjective aspects are emphasised when using projective techniques. This is due to the fact that the information that is being gathered through projective approaches is so highly subjective and hence context-bound. However, this is also a strength of these approaches as it is the rich understanding that a person has of an object, issue, event or state of affairs that is of interest to the researcher, and attempts are not being made to produce object fact but rather to encourage subjective and personal interpretations. It is important, however, that the researcher is aware that the research situation and their[1] presence as a researcher will inevitably be influential and intrinsically tied in with the responses the participant commits. With this said and kept in mind, we will now turn to our consideration of different forms of projective techniques.

Presenting Projective Techniques

As we discussed in the earlier pages, people carry with them lots of information that they themselves struggle to retrieve and comprehend. It is awe-inspiring that simply asking someone to complete a sentence has the power to both unveil and unravel deep-seated thoughts.

Let's test a text-based projective technique.

Sentence Completion:

Finish the following sentences with the first word that comes to mind.

- The Russian invasion of Ukraine makes me ___________.
- My doctor is ___________.

- The clothes she wears are ___________.
- When I think about the male dancer I ___________.
- Immigrants are ____________.
- The Asian student is ___________.

How did you answer those questions? The first thought that comes to mind may hint at immediate biases. Did your answers shock you? Were you tempted to adjust your initial thoughts? Whilst not all questions and answers may provide a sense of shock or breakthrough, they open an opportunity for revealing less encumbered or cognitively corrected thinking. This unencumbered thinking produces knowledge, sometimes for both the participant and the researcher, and leads to raw and authentic qualitative data.

Alright, let's try a technique that is perhaps a bit lighter and is commonly used when conducting research by marketers conducting branding projects.

Brand analogies:

- When you see Mcdonald's what is the first thing you think of?
- What do you hear when you read, "Taco Bell"?
- What is the first word that comes to mind when you think of "Patagonia"?
- What colour do you see when you read the word, "sustainability"?

Your mind is made of countless associations that lay dormant unless called upon. The brilliant frameworks of projective techniques are cloaked as simple exercises but reveal key components of one's psyche. These findings may be as light as determining what colour to use for a company logo, or they may be deeper, unveiling gender and racial bias within sensitive ethnographic studies. We do not aim to tell where to apply such techniques, but we do seek to highlight the power of these stimulating exercises.

Whilst we will not yet dive into research design, we will introduce a series of projective approaches that we list further. As you read through the following chapter, we suggest you consider your own potential research and how these approaches will have been and/or would be of great benefit within these contexts.

The approaches we present include:

Word Association	Personification	Sentence Completion
Story Completion	Thematic Apperception Test	Choice Ordering
Collaging	Brand & User Imagery	Timescape
Tradeoff	Projective Drawing	Family of Brands
Laddering	Guided Imagery	Talk Balloons
Role-Playing	Guided Fantasies	Brand Analogies
Survivor Island	Billboards	Dial Testing
Listing	Grand Tour Technique	Repertory Grid Technique
Ranking	Visual Props	Mini-Tour Technique
Vignettes	Drawing	The Echo Approach
Cartoons	Descriptive Diary	

Let us now start to think of some projective approaches in slightly more detail. Many of the projective approaches we describe will be presented within a consumer psychology or a market research context. This is because many of the approaches have been developed for use in this arena. However, the projectives we describe may all be readily used within other areas of human subject research.

Projective Approaches

Text-Based Projectives

- Sentence Completion: As we have already seen in this chapter, a sentence completion projective approach requires respondents to finish a sentence in any manner that they deem appropriate.
- Story Completion is a verbally based projective in qualitative research. Story Completion takes sentence

completion a step further and asks participants to complete a story. This approach requires participants to focus on their feelings. This methodology is fairly simple and comes with rudimentary instructions. The moderator asks the respondents to tell a story about the product or brand, event or object of interest. Respondents, when asked to tell a story about a picture, product, brand, event and so forth, will describe a detailed tale where they increasingly reveal their own attitudes and feelings towards an attitudinal object as the story unfolds. Unlike some other projective techniques, story completion is undertaken individually, with people usually completing the story on a paper form. An example of this technique is, "When I walk into Thrift City Stores, I get the feeling that ____. I think that their stores are better/worse than their competition because ____. If a friend asked me tomorrow for a recommendation, I'd tell them _____". From this example, it can be seen that story completion does not require a participant to write the ending to a story but rather to add words of phrases at strategically determined places in a short storyline.

- Talk Balloons is quite an imaginative technique that reveals strong insights. For this technique, participants are supplied with cartoons. Talk balloons using cartoons of people talking are ideal for concept testing and branding, and for other situations in which you are attempting to find out someone's reaction to being placed in a situation. Talk balloons leave just enough to the imagination. Respondents fill in the blank, even fill in the speakers' expressions in a cartoon. Talk Balloons are also commonly referred to as the Cartoon technique. Participants are asked to react to a cartoon or complete an empty speech bubble with a word, phrase or thought, which can effectively lead to a discussion: A cartoon may be used to begin a discussion on a difficult topic.
- Survivor Island is a technique for winnowing out a large number of features or attributes from a participant. Ideas

generated from participants are written on stickies and placed on a picture or other representation of an island. After the idea generation stage, ideas get voted off the island or placed on the coastline. More specifically, the ideas that the participant considers as "must haves" are left on the island, whilst the "nice to haves" are located on the beach. Any idea that is thought of as a "save it for later" idea is placed in the ocean.

- The Billboard technique is most commonly used for identifying the essence of a brand, product or some other entity. For example, Billboards don't have too many words, since they are aimed at conveying a message and gaining your attention whilst you are driving by, and they therefore have to keep the message short. In this technique, the moderator prompts the participants to come up with a billboard for a product or service using just three words: a phrase, or three separate words.
- Descriptive Diary asks participants to jot down general thoughts and recordings at times throughout their day or week.
- Listing is a fairly simple technique used in qualitative research that can provide beneficial returns. This technique asks participants to make a list of items, either individually or as a group. This can be followed up by ranking, comparing or discussing the items listed. For example, ask participants to collectively "Make a list of all the water sources in this village" or "List the benefits and limitations of the health screening program".
- Ranking is quite similar to the Listing technique. However, participants are asked to rank items by given criteria (e.g. cost, quality, effectiveness and convenience). For example, a moderator may tell participants "Using the items on the list, rank them from the highest to lowest quality". In this technique, participants may be provided with a pre-prepared list of things to rank, or a list may be generated by the group itself and then ranked.

Verbally Based Projectives

- Word association is a projective technique that requires subjects to respond to the presentation of an object by indicating the first word, image or thought elicited by the stimulus. In less complex terms, the moderator will ask participants what words come to mind when a stimulus word or phrase is presented – for example, Coca-Cola: beverage, thirst, fun, relaxation – respondents reveal valuable information regarding their attitudes and beliefs.
- The traditional definition of personification is the attribution of a personal nature or human characteristics to something non-human, or the representation of an abstract quality in human form. However, in qualitative research, it refers to respondents being given a number of words and pictures and instructed to select those that they associate with a brand or product or any other object or event. The respondents are also asked to explain their choices, providing reasons for their selections or attributions. Personification is also known and commonly referred to as anthropomorphism. An alternative definition of anthropomorphism or personification is linking brands, products or things to characters of people or to famous people. For example, a new kind of yoga mat, the moderator asked participants what celebrity the yoga mat would be. Additionally, if the moderator asked, "If the UK Conservative party were a person, what kind of person would they be? What would their hair be like? Their clothing? What sort of activities would they engage in".
- TAT is projective that can unlock powerful insights. This method provides respondents with a set of T.AT cards where each picture depicts an individual (or more than one individual) in a fairly ambiguous environment engaging in an indistinct behaviour. It is the respondent's task to develop and tell a story of what may be occurring in the picture. This procedure is repeated for several TAT card combinations. When TAT is used within a clinical setting, based upon the sum of the explanations,

the clinician is able to deduce the theme of the patient's stories. This motif provides insight into the psychological conflicts, past and present, central to the patient, thus making it a handy methodology for many researchers to use in research or consumer settings.

- Timescape is an advantageous method in qualitative research. This technique permits participants to explain their view of the future. Often this involves setting the stage with a scenario, such as "pretend you are on a spaceship and you are gone for ten years. When you return to Earth, what changes have you noticed in terms of how people, do their work, communicate, travel, etc.?" Participants' responses are analysed to see how well the present solutions satisfy potential future developments, possibilities or requirements.
- The Family of Brands technique asks participants to describe a brand as a family member in an extended family or an animal in a zoo or an automobile. The set of possible choices may be limited in order to allow the achievement of consensus amongst several participants. This procedure helps to understand values or feelings that a person cannot readily express.
- Guided Imagery simply asks participants to imagine a brand or some phenomenon through taking a journey in a dream-like state. The details of this experience are then evaluated to provide clues as to what participants feel about the target of the imagery.
- Role-Playing is an interactive technique that can produce strong insights. For instance, participants act out different roles within a scenario, for example: clerk and customer, friends chatting about new products or services, strangers in front of a new store, and people discussing a news article. The dialogue the individuals use in the role-playing situation is then analysed and will then reveal hidden emotions and values that may be hard to uncover through direct questioning.
- Guided Fantasies is a technique quite similar to role-playing in the way that it deals with a fictional scenario. The

moderator describes scenes using metaphors, from opening doors to mansions symbolising brands to interplanetary travel to planets representing brands. "You land on Planet Apple. What are the people doing? What do they look like?" It's a way to creatively personified brands. This technique is ideal for branding research.

- Brand Analogies is quite simply prompting associations of established brands across different categories. For example, "If Bernie & Phyl's [local furniture store] were a car, what kind of car would it be?" This may be used outside of the market research context by asking questions such as "If the Green Party was a car, what kind of car would it be?"
- The Grand Tour technique asks participants to reconstruct a significant segment of an experience or of an event. For example, in interviewing a counsellor, an interviewer might say, "Take me through a day in your work life". Or in working with a student teacher, an interviewer might ask, "Reconstruct your day for me from the time you wake up to the time you go to bed".
- The Mini-Tour technique is nearly identical to the Grand Tour technique. However, this methodology asks participants to reconstruct the details of a more limited time span or of a particular experience. For example, an interviewer might ask a vice-principal to reconstruct the details of a particular disciplinary session with a student; or an interviewer might ask a teacher to talk about the experience of a particular conference with a parent.
- The Echo Approach is when the moderator marks down a note while a participant is speaking that he/she/they want to come back to later but do not want to interrupt the participant's thought process or their spontaneous elaborations associated with the current question. After the participant has finished speaking, the interviewer might ask to go back to the words they said and elaborate on these a little more.
- Vignettes are short scenarios that are read aloud to the participants and which are followed by a series of specific

questions. For example, participants may be asked what advice they would give to characters in the vignette, whether the scenario is common in their communities, and how these issues are typically dealt with.

Visually Based Projectives

- Collaging is a technique that allows a participant to create a visual representation of their thoughts or feelings. Participants move beyond the bounds of language in an attempt to more accurately and completely communicate a person's opinions in pictures and words to form an alternative definition as a visual representation of participant perceptions. For example, one can do collages that are "pre-/post-": how did you feel about the company (or any other thing or concept) before . . . ? After . . . ? Or two collages may be assembled, one for different options, for example, one for brand A, one for brand B. Traditionally, people would search through magazines and cut and clip them, but now there are electronic collaging solutions, such as Karma Collage, which requires respondents to perform the creation of their collage on a tablet. Additionally, it is important to ask participants to focus on the rationale they are employing in their creative process.
- Picture Sorts are similar to Collaging but involve sorting through images and finding those that represent brands, companies, products, situations and so forth. Visual Explorer and IconiCards have card decks that the participant uses.
- Brand and User Imagery is a technique that is similar to collaging. However, this technique permits a participant to evaluate the mental disconnections between their perceptions of the brand (or activity) and perceptions of users of the brand (or participants in the activity). For instance, participants are asked to convey their perception of the brand/activity, and then later they are asked to explain who might use the brand or participate in the

activity. Discrepancies can be further probed to provide deeper insights.

- Projective Drawing is a technique that asks participants to draw potential consumers for or of specific products, services and so forth. Participants are also asked to provide basic demographic characteristics and attributes, features, qualities and so forth of these potential consumers. Drawings are then discussed to explore differences and features in these perceptions.
- The Visual Props technique quite simply shows props to the group to stimulate discussion. The materials that are used include educational material (i.e. posters and videos), current events (i.e. newspaper article and media clip), advertising material and products.
- The Drawing technique asks participants to make a drawing on a specific topic given by the moderator. For example, students may be asked to draw a tree and then in the roots of the tree, write all the causes of stress for graduate students.
- Mapping sounds like it might be similar to concept Mapping (see below), but in reality, the approaches are quite different. In this case, when mapping is used, participants are asked to sketch a map of their neighbourhood and then mark certain items or places on the map as directed by the moderator. For example, participants may be asked to sketch a map of their neighbourhood and then mark all the places that they believe are safe or dangerous, hygienic or unhygienic, and so forth.

Mapping Techniques

- Mind Mapping is a powerful methodology in qualitative research. This technique helps generate ideas and makes clear the relationship of these ideas between and within groups. Mind Mapping also derives concepts directly from participants that can be used to communicate product benefits to a broader market. Obviously, as we stated at the start of our writing, whilst this technique has been very widely used in consumer settings, the approach is also very often employed in other research domains that are interested in human activity. When using mind mapping, the mapping exercise may be done at the individual participant level or at the group level. It also facilitates brainstorming and idea generation.

- Choice Ordering is a technique used to force participants to evaluate their input by asking them to rank or order certain factors or attributes associated with an event, object, concept, brand, service and so forth. This exercise is very commonly combined with mind mapping, especially when the object of the mind map is to establish preferences or desires in association with the target phenomenon.
- Concept Mapping is a technique that has the ability to demonstrate how people visualise relationships between various concepts. This is closely related to cognitive mapping from within the discipline of environmental psychology. Concept maps provide a visual representation of dynamic schemes of understanding within the human mind, yet some debate exists about what is and what is not a concept map.

Sort Methodologies

- The triadic sort is good to its name as it provides three items to a respondent as either physical objects or things, or as words or pictures on cards. The participant is then asked about similarities and differences between the three things and is asked how two of these are similar to each other but different to a third. The procedure identifies the underlying understanding that the person has of the three items.
- Pile sorts is a technique similar to triadic sorts. However, in this procedure, participants are asked to sort a range of images, photographs, or words into different piles by various criteria. Participants may be given the criteria by which to sort the items by the moderator or be asked to

develop their own criteria of categories that distinguish items. In the former approach, the approach identifies how participants understand items in terms of criteria that are being investigated and in the latter form of the procedure, sort criteria themselves are being generated as well as items being sorted into criteria.

- In the Q-Sort approach, a participant is presented with a set of statements about some topic, and is asked to rank-order them. The ranking is often in terms of how much they "agree" to "disagree" with each statement, but the ranking may be between any two poles such as positive to negative, like to dislike, cheap to expensive and so forth. Whatever the poles terms used the assessment is in terms of the extent to which the participant understands the criteria to apply to each of the statements.

Other Projective Approaches

- Laddering is a technique that is most commonly used in in-depth interviews. When using this approach, participants are asked to define the attributes of a product, service, thing, concept and so forth, and then describe the rational benefits of that. Exploration is then undertaken by the researcher to elucidate why those benefits are of importance to the respondent. The chain that this procedure develops links the attributes held by products or other things to psycho-social or emotional benefits that are held by consumers and other respondents. The chain can be an upward ladder to an emotional benefit, or can be reversed as a downward ladder to find out the disadvantages of a competitive brand, and so forth, in terms of its rational benefits.
- Dial Testing Using Perception Analysers is a technique that permits real-time analysis of participant feelings towards a given stimulus. The feelings are captured using a numerical dial and may be segmented by age, income, gender and so on. The stimulus may be played back during a focus group with the numerical data displayed as

graphs, and this data then becomes a source for further probing on likes/dislikes and strengths/weaknesses of the stimulus.

- The Trade Off technique gives participants' shapes to fit within a space on a piece of paper. Each shape represents a particular product attribute. However, participants cannot fit all shapes into the space; therefore, to complete their box, participants must trade off certain attributes for others. This prioritisation can then be evaluated at the group level and probed for clarification. A variation of this is a card sort where participants must sort attributes into categories such as must have, nice to have, not important, for example. The card sort technique (is a variation of the sort procedures already mentioned) entails providing a group of users with a set of cards. Written on each card is a concept or piece of information from the set that needs to be organised. Users then sort the cards with similar concepts into piles. The cards are scored and the data is entered into a statistical analysis programme. A statistical cluster analysis can be used to create a composite of all or various groupings of users. The technique is based on the assumption that if users group cards together, the concepts probably possess a meaningful similarity to them which will help in understanding their relationships to whatever it is that is being investigated. For example, the result suggests how users would organise a given set of concepts, which can be very valuable information when organising a system or Web site.
- Repertory Grid Technique is an approach researchers use to understand the relationship between conscious or subconscious constructs that lie behind each single act of judgement that a person makes. The approach comes out of Personal Construct Psychology developed by George Kelly (1963, 1991a, 1991b) and see also Fransella and Bannister (1986, 2003). Personal constructs are thought to underlie how we understand events within our world and the judgements that a person makes in their day-to-day lives. The repertory grid technique has been designed

in order to explore a person's personal constructs, the structure, content and interrelationships of constructs. For example, each of us has many such implicit theoretical beliefs (or what may be called construct systems) about all aspects of our lives, such as billiards, love affairs, money, children or God.

Conclusion

We have finally reached the end of this chapter, and hopefully, we haven't scrambled your brain with all the different techniques and methodologies. In this chapter, we have attempted to familiarise readers with a wide range of projective approaches. We reviewed text-based projectives, which that employ writing or texts to elicit information from participants; verbally based projectives, which involve the use of spoken approaches in their gathering of data; visually based projectives, which offer images of physical items as stimuli to participants; mapping techniques, in which participants make maps or arrange things or concepts spatial; sort methodologies, in which respondents are asked to sort items or concepts into spatial arrangements to represent similarities and differences between these stimuli; and some other projective approaches.

In presenting this wide variety of techniques, the aim of this chapter has been to provide the reader with an introduction to projective techniques that exist within qualitative research. These approaches are valuable to researchers as they help facilitate the gathering of extremely useful information and insights. One of the main, most useful and unusual characteristics of all of the approaches we have provided details about is that they all in some way attempt to break through conscious or subconscious barriers that a respondent may employ in order to protect themselves in some way. Furthermore, the respondents who take part in a project procedure may find it both engaging and easy to participate as the design and nature of projective techniques encourage creativity and expression of attitudes, beliefs or emotions in a safe, no judgement zone.

Hopefully, most participants should actually enjoy completing a projective exercise!

Having introduced you to the wide range of approaches we will consider in this book, in the next chapter, we start to look more closely at specific procedures and we commence with projectives that employ some form of sorting in their methods.

Note

1 Throughout this text, it is the authors' intention to use non-binary gender terms. We hope that the reader will excuse any grammatical obscurity or vagueness that may arise because of this practice.

References

Fransella, F., and Bannister, D. (1986). *Inquiring Man: The Psychology of Personal Constructs*. London: Routledge Kegan & Paul.

Fransella, F., and Bannister, D. (2003). *A Manual for Repertory Grid Technique*. London: Wiley Blackwell.

Kelly, G. (1963). *A Theory of Personality: Psychology of Personal Constructs*. New York: W. W. Norton & Co.

Kelly, G. (1991a). *The Psychology of Personal Constructs: Theory and Personality Vol. 1*. London: Routledge Publishers

Kelly, G. (1991b). *The Psychology of Personal Constructs: Clinical Diagnosis and Psychotherapy Vol. 2*. London: Routledge Publishers

3 Sort-Based Methods

Chapter Summary

In this chapter, we exclusively discuss sort-based methods. We review the broad reasons to utilise these techniques and offer the reader an in-depth understanding of how to execute these approaches. We review why one might employ an open sort versus a closed sort, a single sort versus a repeated sort and qualitative versus quantitative sorts. The approaches we review in detail are pile sorts, picture sorts, triadic sorts, Q-sorts and card sorts. We discuss the ways in which Q-methodology supports and connects qualitative and quantitative research and the benefit this multivariate approach has in one's research. This chapter will also address the vast array of disciplines sort-based approaches may be used for and underline their diversity and flexibility.

Introduction to Sort-Based Approaches

In the first two chapters, we attempted to familiarise readers with the broad range of projective research approaches. In the next few chapters, we consider the different types of projectives in a little more detail. In this third chapter, we turn to deal with approaches that fall under the heading of sort-based methods. The name of the chapter prepares the reader for what to expect from the sort-based procedures that we will consider. That is to say that they all require the participants to sort

DOI: 10.4324/9781003285892-4

some form of material, be this, actual objects, words, pictures or some virtual material.

The reasoning behind the use of sort techniques is one of two major types. In the first of these, the participant is supplied with categories into which to sort the objects, words and so on. On this understanding, the approach simply looks to see under which categories items have been placed. In the second form of sort approach, the participant supplies the categories into which items are sorted and then places items under these categories. The former of the two approaches is interested to test out pre-existing categories amongst a specific set of participants whilst the latter is more concerned with the generation of new understanding in terms of how items are differentiated and understood by participants.

As with all projective techniques, sort-based procedures are varied and flexibly adapted by the researcher to suit a specific research situation. Indeed, throughout this book, we emphasise that projective techniques of all varieties are extremely malleable and the researcher should tailor the technique to provide the specific information they require to answer their research questions. As there are very many ways of sorting things into categories, the illustrations of these techniques that we offer have been chosen by us to provide the reader with an understanding of the various sub-forms of sort procedure rather than constituting the approaches we prefer or advocate. The approaches we will include in our review are as follows: pile and card sort techniques; picture sorts; Q-sorts, and in the following pages, we will address each of these specific types of sort techniques. We also consider the choice that the researcher has to conduct either a qualitative or quantitative search procedure and analysis.

Characteristics of Sorting Techniques

As we have just mentioned, there are several different types of sorting techniques that exist in qualitative research, some of which yield numerical data (which is slightly unusual amongst the projective approaches that we talk about in this book and projectives in general) whilst others are qualitative in their output. The overall defining characteristic of these approaches is that the researcher asks the participant to allocate or sort something into a series of exclusive or non-exclusive characteristics, features or categories. Sorting approaches can be distinguished by their openness, exclusiveness, the sorting task and the character of categories (Fincher and Tenenberg, 2005; Harloff, 2005; Harloff and Coxon, 2007; Rugg and McGeorge, 2005). For instance, a sorting technique that is exclusive in nature, means an item, card or picture, can only be sorted into one category or pile. We talk more about all of the characteristics of sorting procedures in the next section. It is important to understand the various types of sorting techniques that exist and from which the researcher may choose as if the appropriate approach is chosen, they can yield powerful insights into and from the participants. It is also important to note right from the start of this chapter that, as well as providing information in regard to the sorting a participant does, the approaches also allow the researcher to discuss a participant's performance, ask them to elaborate on their sort and to provide much rich and meaningful information. Additionally, sorting techniques may be employed with individuals or in small group settings, as stand-alone procedures or with other approaches, perhaps in the setting of a focus group. Furthermore, sort approaches may be used to produce in-depth insight into why a person categorises the content they have sorted in the way they have or to offer numerical indicators as to the strength of the similarities and differences of perceptions that participants have regarding their sorting preferences.

In what follows, we elaborate on some of the broad characteristics of sorting techniques before we go on to consider the different tests themselves.

Opening Phases in Sort Procedures

The first characteristic that is common across sort-based procedures is the fact that sorting approaches rest heavily on researchers taking time to clearly establish exactly what the specific question is that is being asked in the research and how the research may be designed to produce the answers. At this point, the reader will rightly say that being clear about the questions a project is attempting to answer is required in all good research designs. However, there are multiple reasons why this is particularly true with all forms of projective tests and arguably exceptionally relevant with sort procedures. The first reason for this pertinence is that as projective techniques are very often designed specifically for the situation in which they are used, there is not an already developed test that a researcher may use. Furthermore, due to its subjective nature, the researcher must take steps to ensure that the procedure they use clearly addresses the questions being asked in the overall study's aims. In sort procedures, it is frequently the case that the instructions given to participants are somewhat rudimentary as it is the aim of the researcher to facilitate free expression from those taking part. It is also the case that as there are no direct questions given for a participant to answer, but rather they are presented with cards with words or images printed on them, or physical objects and then asked to arrange these, great care must be taken at the stage of choosing words, images or objects.

Open Versus Closed Sorting

Let's begin with open sorts. In open sorting procedures (Chollet et al., 2014), respondents are instructed to group items or pictures into categories of their choice. In this case, the researcher does not determine the sorting criteria, the number of categories or the labels given to the categories. The reason behind this is to encourage participants to think freely and create their own categorisation structure. Following the completion of the sorting task, participants are asked to elaborate on the categories and names they developed for the categories. This approach is beneficial in identifying and analysing the patterns of categorising and the way in which information is organised by the participant. It can also be useful for researchers when comparing whether participants use the same or different categories for sorting the items. Open sorting is frequently used in a collaborative setting, where participants are able to debate and talk over the approach and the categories everyone within the setting developed. In adopting such an open approach, it is hoped that this will result in the participants coming to an agreement on the categories they developed and the meaning behind them.

Now that open sorts have briefly mentioned, it is important that the counterpart, closed sorts, are discussed. In closed sorts, the researcher supplies participants with a predefined set and number of categories, along with labels or names for them. Participants then sort the items or visually based materials into categories. In some cases, respondents are asked to rank items presented to them, but it is important to note that ranking items in sort approaches is a within-item relational process. This means that items are not assessed individually, but ranked in relation to other items in the specific procedure. For example, college students were asked to rank a series of pictures of golden retriever dogs. They were instructed by the researcher to rank from left (most) to right (least) based on how appealing the visual images of the dogs were to them. The results that this type of procedure produces will not reveal how appealing each individual photograph of a dog is to the participant, rather it will show how appealing they are in relation to the other photographs of the dogs.

Single Sorts Versus Repeated Sorts

Single sorts, as their name implies, are quite simple to explain and understand. In single sort procedures, the participants only perform the sorting technique once. Conversely, we are certain that you can guess that the definition of a repeated sorting is a sorting that is performed several times. In repeated sorts, participants are required to repeat the sorting task either directly after the first sort or after some time has passed. The benefit of repeated sorts is that it provides the opportunity to replicate the approach in order to provide information to help understand the consistency or inconsistency in how people sort items. In repeated sorts, another variation is that respondents aren't always asked to re-sort under the same criteria. It can be instrumental to sort the same items into a new group or pile under a different sort criteria. The repeated sort can then be used to cross examine the results from the first sorting in order to provide a different view or perspective on the categorisation of the items.

Qualitative Sorts Versus Quantitative Sorts

When a researcher decides that a sort-based procedure is a good fit for their research question and topic, they have the option of choosing a sort-based procedure that yields either qualitative or quantitative information. In both of these classes of sort-based projective techniques, the procedures are similar but have essential differences in their aims and procedures. At this point of the chapter, we will not go into details in regard to the ways in which different sort approaches that are either qualitative or quantitative are conducted as this will be covered in the latter section of the chapter. However, before we go into such detail, it will be useful for readers if we describe the main differences between conducting a qualitatively and a quantitatively analysed sort procedure.

An example of what may happen when conducting a qualitative sort is as follows. In this procedure, participants may be requested to sort objects, cards with words of pictures printed on them, into spatial arrangements that represent how they understand the things they are arranging are associated to each other. This procedure is conducted on a flat horizontal or vertical surface. The arrangement is believed to reflect how participants understand the content of the area that is being investigated is structured. The sort may then be photographed or recorded in some other way and the groupings and arrangements of items in relation to each other is noted. The respondent then discusses the arrangement they have produced with the researcher. This procedure reveals the conscious and perhaps some subconscious attitudes and beliefs held by the participant.

When conducting a quantitative sort, the procedure is similar to the qualitative sort method noted earlier. However, in quantitative approaches, the material that is sorted by participants is numerically coded in some way. By attributing a number to a sort item, the researcher is then able to calculate descriptive statistics for groups of respondents. This includes statistics such as means, modes, ranges and variance statistics. By doing this, the researcher can make comments about how typical an individual's response is in comparison to other individuals or to groups of people. If inferential statistics are employed, comparisons can also be made between groups of respondents and the statistical significance of the differences and similarities between respondent groups may be assessed. More sophisticated multi-dimensional statistical procedures (Kruskal and Wish, 1978) may also be used that enable the internal structure of response to be investigated in terms of their psychological nature (e.g. see: Borg et al., 2018; Borg and Groenen, 2005).

Earlier we have provided a brief background to sort methodologies and the general sub-types of techniques that occur within this form of approach. We now turn and discuss selected individual sort approaches themselves.

Details of Selected Sort Techniques

In this section, we will provide details of the following sort techniques: pile and card sorts; picture sorts; triadic sorts; and Q-sorts. Our selection of approaches is not meant to be comprehensive but rather to illustrate the breadth of different sort approaches from which the researcher may select when designing a research project. It should also be noted that sort techniques are all quite similar in their designs but the variations allow for different types of information to be collected in order to better address specific research questions.

Card Sorts

Card sort techniques include the sub-categories of sort tests known as pile sorts or other sorts into other spatial configurations. They also include triadic sorts, picture sorts and Q-sorts, which we will talk about in later sections. Card sorts entail providing a group of users with a set of cards. Written on each card is a concept or piece of information and together all of the cards form a set that contains a coherent body of information in regard to the research being conducted and the overall research questions being asked. It is the task of participants to organise the set of cards they are given. As we described in an earlier section, the sorts may be open or closed. Users are given the set of cards and asked to sort the cards in terms of how the cards fall into specific categories that are either provided to the participant or provided by the researcher. If card sorts are used qualitatively the groupings of cards are noted for each respondent and each participant's reasons for sorting items as they have done is discussed them. If the procedure is used quantitatively, the cards are scored or numerically coded in some way and the data is entered into a statistical analysis programme. Several different statistical packages are available and may be used to reveal the statistical similarities present in the data that arises from a sort procedure. For example, a statistical cluster analysis can be used to create a composite analysis of how the group of respondents positioned all test card items. The output from such an analysis takes the form of a print out that positions the concepts on each of the cards so that a concept is positioned near to other concepts that were rated the same but more distant from the concepts that were rated differently. Highly correlated concepts thus form clusters. The technique is based on the assumption that if users group cards together, the concepts possess some form of similarity. The result suggests how users organise a given set of concepts, which can be very valuable information when organising a system or website. An example of a card sort is shown in Figure 3.1.

Figure 3.1 Illustration of Card Sort

Example:

Participants are instructed to sort nine cards with images on them into the researcher supplied categories of Very Positive (5), Slightly Positive (4), Neutral (3), Slightly Negative (2) and Very Negative (1). Once a participant has completed this sorting task, the researcher enters the numerical scores (one of the numbers from 5 to 1 given above in parentheses) and uses a statistical programme to analyse this data. This procedure allows for composite analyses of the area that is being investigated and to understand participants' perceptions of the similarities and differences among the images.

Pile sort (Trotter and Potter, 1993) is a specific technique which falls under the umbrella of card sorts in which participants are asked to sort a range of images, photographs, or words printed on cards, into different piles by various sorting criteria. During this procedure, participants may be supplied with the criteria for sorting items by the moderator or they may be asked to develop their own criteria of categories that distinguish items. The analysis of pile sort methods may combine qualitative and quantitative techniques in their analyses. Pile sorts begin with attributes being developed by the person conducting the research project, and attributes are then listed on a set of cards, pictorial items are developed or physical objects are selected for the sort. Next, participants are instructed by the moderator to carefully read through all the cards and then make their best judgement to place each card or item into a "pile" or group that they believe best describes the card's content.

In most research studies where pile sorting is used, the cards that a participant perceives to be the most similar are sorted first. Then, the cards with less similarity are sorted to conclude the technique. This process allows participants to sort through a large pile of cards until there are two or more clusters of cards. Alternatively, the methodology can be conducted in the opposite direction. For instance, all the cards are separated into individual piles and the participant then sorts them into one pile. Once the sorting by participants is completed, the researcher(s) calculates the "distance" between each of the piles of cards by measuring in centimetres the actual distances. The results are then translated into a distance or proximity matrix. The matrix is then statistically analysed using some form of cluster analysis (a form of multi-dimensional statistical analysis) with the goal of investigating and revealing the hierarchical structure (hierarchical analysis, see Garson, 2014), or other forms of clustering relationships to help researchers develop a "map" that will demonstrate why and how participants sorted the cards into piles. In this way, the researcher is able to assess and understand how the participants think and feel about the area they are investigating.

It may be useful to provide an example of this approach. For example, a retailer is interested in understanding how their customers perceive their products and also in identifying areas in which customers believe there to be a gap in the products they sell. In this situation, a sample of customers could individually be given either sample of the products available in the shop or cards with pictures or names of the products printed upon these. The participants would then be asked to sort items into piles and to arrange the piles so that the position of the piles reflected the relationship that a pile had to other piles (this is a simple qualitative version of this procedure). Participants are then asked to name the piles they have formed and to comment upon their content items and how they are similar to other items in the pile. In this way, there may develop gaps in the arrangement where none of the products sold in the shop fit. Such a procedure would then help the company to understand how customers perceive their retail offerings and also to identify target areas (the physical gaps in the spatial arrangements) that they need to source and offer in order to provide shoppers with a more complete choice of products.

Picture Sorts

It will be immediately apparent to readers that we have already spoken several times about sorting pictures. However, it may be helpful if we consider these in a little more detail. When a researcher uses a Picture Sort, their main goal is to identify how individuals organise and arrange visual items. Instead of cards with written phrases, words or statements; pictures are used for sorting. The results of the sorting provide the researcher with a form of visual data. The visual data and the observed sorting patterns are then examined and deciphered by the researchers in order to uncover the reason behind the categorisations into which items have been allocated. There is obviously a lot of interpretation that is going on, both by the participants and by the researcher. This level of subjectivity may be seen as being both a strength of the procedure and a weakness. It is a good technique to use with participants who do not find it easy to communicate using written language. Perhaps because of the high level of subjectivity involved in interpreting picture sorting exercises, this is not a commonly used research technique in visual and qualitative research. This may be seen as surprising, especially since studies involving sort methodology are excellent for revealing the latent categorisation of visual items.

In qualitative research studies using picture sorts, a set of pictures or cards containing pictures is presented to the participant. Depending on the preference of the researcher, and the specific questions in the research project, these images can be either printed out or digitally displayed for the informants. Then the researcher instructs the participants' to group the pictures based on the parameters of the study.

Image-based sort methodologies can be an extremely useful incredible tool for researchers to have in their back pocket as these are able to transcend the limitations and restrictions of a verbally based method (Buckingham, 2009, p. 633) and present respondents with the opportunity to convey and demonstrate their thoughts, feelings and perspective in a non-verbal manner. In addition, sorting techniques are exceptionally useful and beneficial regarding time and resources. For example, sorting techniques are usually completed quickly with very little cognitive effort (Stephen, 1985; ten Klooster et al., 2008). One could argue that the lack of including verbalisation in a technique can be advantageous in qualitative research because studies frequently require participants to elaborate on topics that may be particularly challenging for participants to verbalise (Lobinger, 2016; Lobinger and Brantner, 2015, 2016).

Tridiac Sort

As mentioned earlier, triadic sorting is a useful technique that grew out of personal construct theory (Kelly, 1955, 1963,1991a, 1991b) and which has become more widely used in qualitative research as it helps identify distinctions that provide the structure and formation of how individuals categorise information. Often, there is a fine distinction between how we understand the objects, events, states of affairs, and so on that make up an area of research interest and researchers attempt to not impose how they understand the area or subject matter they are conducting research upon. In triadic sorts, respondents are presented with examples of significant aspects of a research area and asked to divide up this content in way that is meaningful to them: triadic sorting is an approach that helps uncover these distinctions.

When starting to conduct a triadic sort, the researcher develops a series of printed cards with words or phrases on them. Together, the cards cover all of the major aspects or concepts associated with the research area. Participants are then presented with three cards that have been selected from the card set and they are asked how two of the cards are similar to each other but different from the third? This is noted and also the respondent is asked how they are different and similar and this too is noted. This procedure is repeated with different permutations and combinations of three cards from the set until the reasons that participants give for their choices become repetitive. This stage is called saturation.

In an example of a triadic sort, fifteen cards may be produced that have the names of different models of cars written on them. Participants would then be randomly presented with three of these cards and asked to speak about the three stimuli cards and to say how two of the cars were similar to each other yet different from the third car. The participants might respond with, "Two of the cars are red and the thirds is blue". The procedure is then repeated with three different cards and the response may be, "I used to own those two but I have never had one of those", the procedure is repeated several more times until the comments become repetitious. The elicited responses that participants give through the triadic sorting technique can yield powerful insights onto how individuals think, feel, how they have interacted and so on, with cars, and any other thing, concept, event and so on, and as in the car example, the insights produced may include many different perspectives and forms of evaluation.

Q-Sort

Q-Methodology (McKeown et al., 1988, 2013) (commonly referred to as Q-Sort) was developed by William Stephenson in the 1930s (Stephenson, 1935; Stephenson, 1953). He had doctorates in both physics and psychology (Brown et al., 2008, p. 722) and his background may be seen in his systematic way of conducting research through the approach he developed. Stephenson wrote a letter to *Nature*, in which he proposed the application of factor analysis (a firm of multi-dimensional statistics) in which the factoring was of individuals rather than their traits. Following his synopsis of the traditional use of factor analysis, Stephenson put forward the idea that it is the people who are being assessed who are themselves involved in a form of assessment of the testing situation. For instance, if a group is given a series of tests of various kinds the individuals being assessed subjectively understand the tests they are given. What is meant by this is that when giving tests to respondents, the tests comprise the content that has been designed into the test, plus the subjective understanding each participant has of the test and the tests yield large amounts of subjective interpretation. From these methodological claims, the notion was advanced that individuals assign meaning to stimuli they are presented with. Moreover, if the stimuli presented are intentionally ambiguous, the meaning that respondents impose will come to the fore and researchers will be able to gain entry into the participants' thoughts and feelings, conscious and perhaps also subconscious, on any topic.

Stephenson's letter to *Nature* and the publications that followed brought about a new, comprehensive and innovative methodology that is extremely versatile. Q-Methodology is now used in several disciplines, including but not limited to, psychology, sociology, policy sciences, discourse analysis, political science and aesthetics. Since its development, Q-Methodology and Q-based studies have been significantly expanded. Below, we will discuss the purpose of Q-Methodology, its techniques and the benefits of studying human behaviour. We will also dive into the methodological considerations and the technical and procedural aspects of using Q-Methodologies in a research setting.

Q-Methodology is the process of studying participants' viewpoints. One of the most commonly used techniques used in Q-methodology is the card sort. The primary purpose of Q-methodology is to investigate respondents' perspectives and behaviours using quantitative psychometric and operational principles in tandem with statistical procedures of correlation and factor analysis statistics. This provides researchers with the tools to conduct a quantitative procedure to yield results that may be able to offer some understanding of participants "subjectivity" or personal point of view on any specified topic. Q-based projects typically consist of a small number of participants, and in-depth interviews of single topics or cases are not out of the ordinary. Due to its often small scale, Q-methods make it possible to conduct studies when funds and resources are limited. Only a basic understanding of Q-methodology and statistical principles is needed to design a Sort study. Education on factor analysis is strongly recommended but not required to know before conducting a Q-based study or attempting to understand a study done by someone else. The fundamental approach used in a Q-Sort approach study involves a participant being presented with a set of statements about some topic, and is asked to rank-order them (usually from "agree" to "disagree").

In the paragraphs that follow, we will take you through a step-by-step process for creating a precise, detailed and creative Q-sort approach.

Step 1: Defining and Building Your Concourse

The "concourse" is defined as a set of all of the things that are thought or said about the topic, item, theme and so on that is being investigated. This stage therefore involves gathering the existing opinions, beliefs, arguments and experts' thoughts on the topic of research interest. The material that is collected at this stage is considered to be the raw material for the Q-sort approach. Essentially, it is necessary to gather the subjectivity of what people are saying and what they understand about the topic. This is an extremely important phase in a Q study, as it is in all sort-based studies as the items that are gathered at this stage in effect set the parameters to the study that follows.

Step 2: Development of the Q Set

A Q set is "a purposive selection of statements" (Brown et al., 2008, p. 2) drawn from the concourse that will be presented to the participants. This is commonly referred to as the Q-sort or Q sample. The assembled set often consists of forty statements, but it is possible to conduct the technique using more or less statements than forty. In order to make the study more productive, it is important for the researcher or the person conducting the study to carefully consider and choose the statements so as to cut out any repetitiveness present between statements. The statements then undergo final revisions and are randomly assigned a number. The statements and its corresponding number are then both printed out, and this forms the Q-deck, which will be used by the researcher when they carry out the procedure.

Step 3: Selection of the P Set

The third step in the process is the selection of the P set. This step is possibly the easiest of the entire process. Essentially, all one needs to do is select a number of individuals willing to participate in the study. All that is required is to have enough participants to determine the existence of a factor in order to compare that to other factors. Please keep in mind that the P set is typically smaller than the Q set. For instance, if you have six participants trying to establish a factor, it is common to have them defining two to five viewpoints. That is not to say that it isn't possible to do more, it is just that it is rarely the case when conducting a Q sort to have a P set that is larger than the Q set.

Step 4: Q Sorting

We now move on to Step 4 in the process, Q sorting. At this stage, the Q set of numbered cards, this is the Q deck discussed earlier in step two, is presented to the participant in a pseudo-random order. After they have been presented with the card

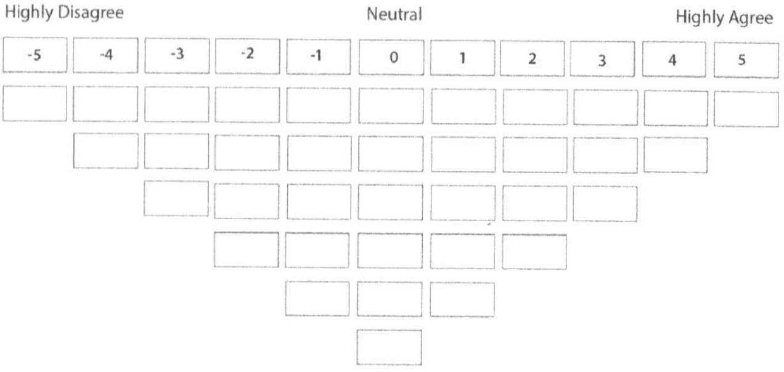

Figure 3.2 Q-Sort Test Sheet

deck, they are asked to rank the statements according to some rule. This is known as the condition of instruction, which is typically the respondent's perspective or point of view on a certain topic or issue. Participants are usually provided with a score sheet and a suggested distribution for the Q sorting assignment. The score sheet is a continuum ranging from one pole of a particular construct to another pole of the construct. For instance, "highly agree" on one end and "highly disagree" on the opposite end (e.g. see Figure 3.2).

Conclusion

Sort-based approaches serve a large number of purposes for a broad array of disciplines as they may be used to gather many types of information that can provide insight into participants' conscious and subconscious motives. Due to their use of ordered arrangements, sort methodologies are able to bridge the quantitative and qualitative divide by yielding data that can be analysed through diverse and flexible approaches. This chapter aimed to equip readers with a general understanding of how these sort-based approaches function and just as importantly why they might be advantageous additions in their research work. Despite Q-methodology's long history and ability to be used in various scenarios and disciplines, it is sometimes overlooked and relatively unused. In this chapter, we discussed the characteristics of sort-approaches, open versus closed sorts, single versus repeated sorts, qualitative versus quantitative and several different types of sorting techniques. Our hope is that our writing provides the reader with an insight into the value, benefits and strengths of using sort-based approaches in research studies.

References

Borg, I., and Groenen, P.J.F. (2005). *Modern Multidimensional Scaling: Theory and Applications (Springer Series in Statistics)*. New York: Springer.

Borg, I., Groenen, P.J.F., and Mair, P. (2018). *Applied Multidimensional Scaling and Unfolding (Springer Briefs in Statistics)*. New York: Springer.

Brown, S.R., Durning, D.W., and Selden, S.C. (2008). Q Methodology. In G.J. Miller and K. Yang (eds.), *Handbook of Research Methods in Public Administration* (2nd edition). Boca Raton: CRC Press.

Buckingham, D. (2009). '"Creative" visual methods in media research: Possibilities, problems and proposals', Media. *Culture & Society*, 31(4), 633–652.

Chollet, S., Valentin, D., and Abdi, H. (2014). Free Sorting Task. In P.V. Tomasco and G. Ares (eds.), *Novel Techniques in Sensory Characterization and Consumer Profiling* (pp. 207–227). Boca Raton, FL: Taylor & Francis.

Fincher, S., and Tenenberg, J. (2005). Making Sense of Card Sorting Data. *Expert Systems*, 22(3), 89–93.

Garson, G.D. (2014). *Cluster Analysis*. Asheboro, NC: Statistical Associates Publishers.

Harloff, J. (2005). Multiple Level Weighted Card Sorting. *Methodology*, 1(4), 119–128.

Harloff, J., and Coxon, A.P.M. (2007). *How to Sort. A Short Guide on Sorting Investigations*. Retrieved from http://methodofsorting.com/HowToSort1-1_english.pdf.

Kelly, G. (1955). *The Psychology of Personal Constructs*. New York: Norton.

Kelly, G. (1963). *A Theory of Personality: Psychology of Personal Constructs*. New York: W. W. Norton & Co.

Kelly, G. (1991a). *The Psychology of Personal Constructs: Theory and Personality Vol. 1*. London: Routledge Publishers

Kelly, G. (1991b). *The Psychology of Personal Constructs: Clinical Diagnosis and Psychotherapy Vol. 2*. London: Routledge Publishers.

Kruskal, J.B., and Wish, M. (1978). *Multidimensional Scaling*. Thousand Oaks, CA: Sage Publications.

Lobinger, K. (2016). "Creative" and participatory visual approaches in audience research, in S. Kubitschko and A. Kaun (eds.), *Innovative Methods in Media and Communication Research*. Wiesbaden: Springer International Publishing, pp. 293–309.

Lobinger, K., and Brantner, C. (2015). In the eye of the beholder: Subjective 20 views on the authenticity of selfies. *International Journal of Communication*, 9, 1848–1860.

Lobinger, K., and Brantner, C. (2016). Different ways of seeing political depictions: A qualitative–quantitative analysis using Q methodology. *Communications*, 41(1), 47–69.

McKeown, B., Thomas, D., and Thomas, D.B. ([1988] 2013). *Q Methodology*. Thousand Oaks, CA: Sage Publications.

Rugg, G., and McGeorge, P. (2005). The Sorting Techniques: A Tutorial Paper on Card Sorts, Picture Sorts and Item Sorts. *Expert Systems,* 22(3), 94–107.

Stephen, Timothy D. (1985). Q-methodology in communication science: An introduction. *Communication Quarterly*, 33(3), 193–208.

Stephenson, W. (1935). Correlating Persons Instead of Tests. *Journal of Personality*, 4(1), 17–24.

Stephenson, W. (1953). *The Study of Behavior: Q-Technique and Its Methodology*. Chicago: University of Chicago Press.

ten Klooster, Peter M., Visser, Martijn and de Jong, Menno D. T. (2008). Comparing two image research instruments: The Q-sort method versus the Likert attitude questionnaire. *Food Quality and Preference*, 19(5), 511–518.

Trotter, R.T., and Potter, J.M. (1993). Pile Sorts, A Cognitive Anthropological Model of Drug and AIDS Risks for Navajo Teenagers: Assessment of a New Evaluation Tool. *Drug and Society*, 7, 23–39.

4 Mapping Approaches

Chapter Summary

In this chapter, we turn our attention to projective approaches to research that employ some form of map or mapping technique in their procedure, or which involve the use and creation of maps of concepts or ideas, that are pertinent to participants. We will present these projective approaches under three major categories: Mapping, Mind Mapping and Concept Mapping. This chapter also addresses ways to plan research utilising a mapping approach and how to collect, analyse and present the related data. We also aim to point out both the pros and cons mapping approaches have so that you are able to make an educated decision on how to include these approaches into your work.

Introduction

In his book, *Representing Place: Landscape Painting and Maps* (Casey, 2002) Edward S. Casey describes the term "placescape" to mean a representation of an area (land or sea) that embodies both the place and the feelings that a person, or people, have of that location. This is an apt term within the context of this chapter's content: mapping approaches. In these techniques, maps are used as *placescapes* of localities and also of other events and phenomena, both physical and mental.

DOI: 10.4324/9781003285892-5

Maps are graphical or diagrammatic representations of places, events or phenomena that exist as representations of the real or imaginary world. Maps are things with which we all have familiarity, be it with printed road and street maps, atlases, or with Google or other maps that we regularly employ on our mobile devices. Maps portray larger phenomena in terms of their useful smaller features. That is to say, a map, say of a city, will portray this larger area (the given city) in terms of its smaller features (roads, railway tracks, rivers, buildings and other features that are present within the city). These smaller sub-components of the city are useful to the reader of the map in that they assist them in understanding or navigating the city or other region being mapped. Maps represent features that we can perceive and to which we may develop cognitive or affective attachment. Thus, a map is a presentation of data or information in a graphical form that shows the spatial arrangement or distribution of sub-features of a place or other phenomena across the area being mapped. The common features that all maps possess are that they are representations of something other than themselves and that the information that they provide to the person who is viewing the maps will help them to better appreciate and orient themselves in relation to the phenomena or event that is being mapped (see MacEachren, 2004, for details about the ways in which maps operate psychologically).

Within social research a researcher is often interested in understanding the specific form of map that a person creates in their mind and carries around with them, consciously or subconsciously, to many situations. The social science researcher may also interested how such psychological maps are created and their interest may extend from mental representations of physical locations to how such abstract and non-tangible phenomena, such as the items on a menu, may be represented in a map that reflects the relationships that a person believes to exist between these food items.

The term that is used to describe these forms of psychological construction procedures is mapping. This is a verb and indicates that the person is actively involved in the process of creating a map of something rather than responding to an already-extant diagram of a phenomenon or an event. Thus, the participant who is completing a mapping procedure is usually required to record, in an appropriate degree of detail, the spatial distribution of the event, concept, phenomena and so on, with which the research is concerned. It should also be noted at this point that if a researcher is concerned with the design of a physical location, for example a shopping mall, then the map the participant will be asked to produce will indeed be something that approaches what we understand as a location map. However, if the subject matter of the research does not concern itself with a physical location, then the map produced will not be of this familiar form. For example, if a person was asked to produce a map that represented the relationship between a selection of confectionary goods, the spatial arrangement that they produced would reflect their attitudes and opinions about the different items rather than any physical "real world" distribution of the items.

As the maps that a respondent produces are personal mental phenomena, they are called cognitive maps (Downs and Stea, 2018). These forms of maps are visual depictions of how a process or concept is understood to exist and presented as a visual metaphor of the relationship between parts of the phenomena of interest in the specific research project. Cognitive maps do not have rules that govern the relationship of items within the map in the way physical maps have to reflect actual physical or geographical spatial relationships. Instead, the relationships represented in a cognitive map are determined by the respondent. The notion of the cognitive map has been enlarged to encompass map-like mental representations by authors such as Tony Buzan (2018).

Having introduced the notions of mental maps and mental mapping as, respectively, representations and the processes of deriving such representations, we will now turn to some specific forms of mapping techniques that are used in social and psychological research.

Mind Maps

Mind maps are well suited for qualitative research studies due to their versatility. For instance, mind maps can be used to plan a project, collect and analyse data, and present findings. If we consider for a moment that different people learn and compartmentalise knowledge differently. Furthermore, people think and comprehend words, graphics and images in different ways. For qualitative researchers, using focus groups and in-depth interviews as the sole technique for collecting data may be relying on psycho-linguistic assumptions about the role of syntax, semantics, and context to guide their construction of meaning (Cassirer, 1946). Authors have put forward the idea that people live two lives, one in their head and the other as part of a social construct (Habermas, 1976). What is meant by this is that consciousness is something that people experience both internally and through their interactions with others (Husserl, 1970). Mind maps provide a potentially less cognitively controlled way for research respondents to convey and portray their experiences, their understanding of events, their likes and dislikes, their pleasures and their difficulties, and so forth. When participants create a mental map, this may be a pleasurable and non-arduous experience that may help to avoid guardedness by participants and the expectations and assumptions built into language and questioning (Korzybski, 1933). Mind maps provide a unique strategy to break away from the conventional limitations of presenting experiences (Hathaway and Atkinson, 2003). It should be remembered that in most mental map procedures, the researcher is able to supply many of the stimulus terms that are the basis of the map, or elicit these from participants. In the former case, the maps produced will be consistent, address a constant theme and be comparable. In the latter instance in which all terms in the map are elicited from the participant, the results will be more idiosyncratic and it is unlikely the maps made by different respondents can be directly compared and contrasted with each other. It is likely that you will be better able to understand this statement after reading the rest of this chapter.

Mind and mental maps can be of use to researchers in many different ways and help them in both the planning and execution of a research project. In the following pages, we consider some of these potential uses.

Planning Research

Mind maps can be helpful when planning a qualitative research project. By first outlining the entirety of the research project, researchers are likely to benefit from being able to see the various steps, tasks and activities laid out as a map in front of them. For example, a researcher's project outline may consist of a list of participants, data collection design, conducting projective techniques and other data collection methodologies, analysing data, and writing up and organising data to be presented. Thus, making maps an effective tool for planning research projects.

Collecting Data

Whilst project management and planning is an extremely useful way in which to employ a mapping procedure, the main use for mind maps is collecting data from participants. In the next section of the chapter, we will take you through an illustrative example of how to develop your own mind map. By doing this, it is hoped that those new to the procedure will quickly be able to grasp and appreciate the fundamental approach behind mapping procedures.

To make your own, first take a pen or pencil and a blank piece of paper, as large as you have, and turn the sheet horizontally. To commence, think of a topic that you find interesting. For example, think of a news headline, a sports story, some local or personal issue or your favourite food. In the centre of the page, draw a circle that is about 5 cm in diameter and then write the topic (a word or phrase) you have chosen in the middle of the circle (if you wanted to take extra time and effort, you could use an image instead of a phrase, either a photograph or a sketch you have produced yourself). As an example, let us say that the topic you decided upon was cricket. Now, draw a line out from the edge of the circle that is about 10 cm long and ask yourself, "when I think about this topic, what is the first thing that comes into my mind?" Write this word, phrase or add another image at the end of the line you have drawn. In the example of cricket, I may have come up with the phrase "different forms of the game". Repeat this step from three to five times. In terms of cricket, you may, for example, as well as "different forms of the game", have identified the sub-aspects of the media through which you watch a game, the women's game, facilities at the grounds and social aspects.

At this stage, you have constructed the skeleton of a mind map that looks rather similar to the sun, with a centre circle with radiating ray-like lines emanating from this. As we turned our sheet of paper to be in a landscape orientation, it is a good idea to write more on either side of the centre circle and less above and below this. Now, go back to your first sub-phrase and ask yourself, "when I think of this word or phrase in the context of the concept written at the centre of the map, what do I think of?" In our example of cricket, the first sub-phrase was "different forms of the game" and when thinking about these, you may have come up with: test cricket, T20, village green, four-day match, the Hundred, one-day international (ODI). Write each of these words or phrases around the sub-phrase "different forms of the game". Having completed this, repeat this with the other words or phrases you wrote around the central topic.

What you have done is to create a mind map. You can finish a mind map after the first ideas have been written around the centrally positioned concept, or you can undertake a second round of differentiation, as shown in our example of cricket. If you have a large sheet of paper, or if you perform the construction activity on a white- or black-board, then you may wish to perform further rounds of differentiation of the divisions that are generated. Two examples of mind maps are provided in Figures 4.1 and 4.2.

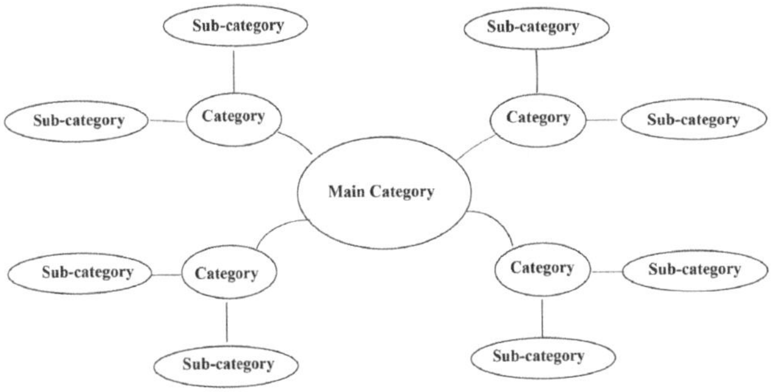

Figure 4.1 Example of Mind Map #1

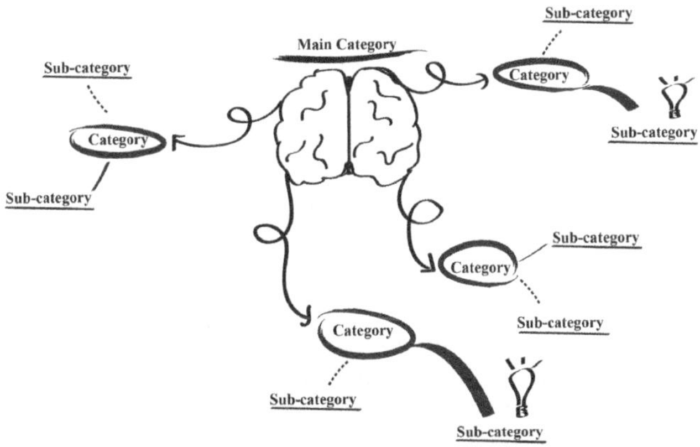

Figure 4.2 Example of Mind Map #2

When using this approach in a research context, at whatever stage the researcher decides to finish the creation of the mind map, the participant should be asked to elaborate on the words and phrases they supplied. For instance, with our example of cricket, the participant may be asked, "can you tell me more about what *test cricket* (for example) means to you".

Participants should be encouraged to be as creative and unique as possible when they are constructing their map and to think about the words or phrases from different angles. By doing this, the researcher avoids participants by simply producing a description that contains little personal meaning: the discovery of what a topic means to the person being the aim of the approach. It is also possible to make the main map more intricate by, for example, using thicker lines to show a stronger connection to the main image or word compared to a thinner line. In addition, you may consider using various colours to help organise the map and to express the importance of different words and phrases. You may also draw linking lines across the map if cross-linkages become apparent between words and phrases in different word clusters and ask the participant if they can label what this connection means to them.

Having gone through the process involved in creating a mind map, we will now turn our attention to how the information that comes out of such a procedure may be analysed.

Analysing Data

Once a mind map has been completed, the information present within this must be analysed and interpreted within the context of the study. Analysing mind maps is different compared to analysing interview transcripts or other forms of data collection. As with all qualitative research approaches, mind maps yield information that requires a considerable degree of interpretation on the part of the researcher administering the procedure. When it comes to analysing mind maps, researchers read and analyse the completed maps. As noted earlier, the maps produced also offer the opportunity for the researcher to ask more and probe the person completing to provide more information about their maps and the probe for greater meaning. The use of maps may draw out information from the participant that they may be unwilling or unable to verbally articulate in other approaches. The mind map is, by its nature, a procedure that allows the flexibility to respondents to form new branches and to develop new connections and relationships between concepts (Tattersall et al., 2007). When analysing a mind map, the researcher is able to fully immerse themselves into the data of the mind map so that they can identify key concepts, themes and patterns. The process that a researcher engages in when analysing a mental or mind map is similar to the constant comparison method (Glaser, 1965) that is widely used in other forms of qualitative data analysis.

Other Types of Mind Maps

In this chapter, we have presented one type of mind map, but there are many other forms and you may also be able to creatively design your own form of mapping procedure. The example of a mind map that we have used is sometimes called a spider map or web, as it is similar in design to a spider's web. However, other designs exist and may be employed. An example of one other format is a concept map known as a semantic map.

Semantic Maps

The use of a semantic map is another way in which a researcher may attempt to delve into the understanding a research participant may have that is of interest to their research project. As with other mental mapping procedures, the creation of a semantic map by a respondent avoids their needing to articulate their thinking behind the creation of a semantic map. However, as mentioned earlier in this chapter, map making is a very useful way in which the researcher may get a participant to "open up" and describe the thoughts and feelings they have that are associated with the map they have created.

A semantic map is similar to the previously mentioned mind map and much of the procedure for creating and analysing this type of map applies in this instance too. However, a semantic map has distinct characteristics such as linearity existing in the concepts mapped and that concepts possess two opposing poles that either the participant or the researcher identifies. We will now briefly outline the procedure for using this technique.

The procedure starts in a similar way to the creation of a mental map in that the researcher presents the participant with a large sheet of blank paper, or a white- or black-board. The researcher then draws a vertical straight line from near the edge of the top of the sheet to near the bottom of the sheet. Enough space should be left to allow a word or phrase to be written between the end of the line and the edge of the sheet. This procedure is then repeated by the researcher drawing a horizontal line across the sheet. Arrow-heads are added to the ends of the lines pointing towards the edges of the surface. Each of these lines represents a construct that is pertinent to the research domain that is being researched and which the semantic map will represent. By construct, what is meant is a meaningful dimension that a participant uses to understand the domain being mapped. The next step therefore involves the researcher either supplying a name for each of these constructs or eliciting this from the participant. The choice of the construct is probably the most important in determining the success of this method in answering the questions that you have in your research. As mentioned about other qualitative research methods, if the constructs are labelled by the researcher this will allow comparability between respondents and may ensure that research questions are directly addressed. However, by supplying the construct names, the researcher is imposing what they believe to be the important constructs rather than allowing a respondent to have free expression.

Let us imagine for a moment that a researcher has a specific project that is concerned with trying to understand voting preferences in an upcoming election. The major issues that the press have touted as being the issues that differentiate the candidates, and which will decide the results, are LGBTQ+ issues and the economy in terms of whether a candidate supports increasing or cutting public spending through tax raises. In this instance, a semantic map could be used to see how potential voters see the candidates in terms of these issues. The researcher therefore prepares a semantic with the two straight arrow-headed lines drawn on this. At the top of the vertical line they write "for LGBTQ+ rights" and at the bottom, they write "against LGBTQ+ rights". On the left-hand side of the horizontal line, they write "against raising taxes" and on the right-hand end of the line they write "for increasing taxes".

What we now have is a sheet of paper, or other surface, that has a cross on it that is labelled to represent the two political issues. Let us imagine that there are eight candidates standing in this constituency. The researcher then asks a participant to look at each of the eight candidates' names and to think about their positions on LGBTQ+ rights and taxation and to locate them somewhere on the semantic map. To illustrate this, if the person looks at the first candidate and believes them to be strongly for LGBTQ+ rights and to strongly advocate raising taxes, then they would write the candidate's name towards the top right-hand corner of the map. If the second candidate advocates the opposite positions on these two issues (against LGBTQ+/against tax raises), they would write their name at the bottom left of the map. The third candidate who is strongly for LGBTQ+ rights but slightly against raising taxes would be located near the top of the sheet and slightly to the left of the vertical line. The other five candidates would occupy positions determined by the two interacting constructs.

The procedure is the same for creating a semantic map when the constructs are elicited from the participants rather than being supplied by the researcher. In this situation, an initial phase to the map creation is required during which the researcher has to establish the important constructs for each participant. In reference to the voting example given earlier, this would require the researcher to carefully talk with the respondent and ask them, "when you think about the upcoming election, what do you think are the most important issues". Instead of directly asking this question the researcher may take some time to discuss this and ask several questions and discuss the options before the participant is asked to decide on the two main issues. When this stage is complete, the rest of the procedure is as has already been noted. Analysis is also similar to that undertaken with earlier mentioned approaches.

So far in this chapter we have presented several different types of mapping techniques that a researcher may decide to use in their research. In order to assist in this decision, in the section that follows, we will list some of the pros and cons of these approaches.

Mind Maps Pros and Cons

When discussing mind maps in qualitative research, it is of utmost importance to mention the values, limitations and challenges. To begin, we will discuss the advantages of mind maps and other forms of mapping procedures. Mind maps can provide researchers with and organisational structure for a large mass of information that may not be easy to digest when this is presented as a table or in prose. Additionally, the output from mapping procedures allows researchers to view participants' ideas, thoughts, feelings, their behaviours and the relationships between these. Mind maps may be beneficial for researchers in several different ways. Most importantly, they help elucidate and explicate the perspective held by respondents. Researchers are given the opportunity to identify and understand respondents' points of view, standpoints, attitudes, beliefs and how they conceive the relationships to exist between concepts and sub-aspects of a concept. Another benefit that mind maps supply is efficiency in their ability to solve both applied and theoretical problems. Mind maps are a relatively quick methodology to deploy is not overly time-consuming to complete and can therefore offer rich insights in a short amount of time. In addition to being time efficient, mind maps are cost-effective. When used in its most basic format, all that is needed to conduct a mental mapping procedure is a pen or pencil, a piece of paper and someone willing to participate in the study. Mind maps promote brainstorming by the participant and facilitate the effective communication of insights into the personal meanings an individual has, in a manner that can enable the researcher to learn a lot about the participant and how they think. Time efficiency is an important advantage as this allows the allocation of time to elsewhere or to run supplementary and follow-up mind mapping sessions.

As well as having advantages, it is important to note that there are also disadvantages associated with using mind maps in a qualitative research study. One of the main disadvantages of mind maps is that it does not have any limits or restrictions. This can be seen as a disadvantage because an infinite number of categories and branches can be created by a participant. The issue behind this is that many maps can get too large causing confusion or for valuable information to be lost. Similarly, if they are used in their most open or unguided by the researcher format, mapping may produce idiosyncratic data that may not answer the questions the researcher originally posed.

In this chapter, we have considered how researchers may use a technique that asks a respondent to produce some form of mental map of a specified content area. Below we will consider one final use of mental maps: the presentation of information.

Using Maps to Present Information

Whilst not actually being a research technique or approach, we feel it is worthwhile mentioning the use of a form of mind map to present the information that is generated in a research study as many researchers reading this book may use this technique. Mind maps can also be beneficial when presenting the data that has been collected in a research study. This effort is to streamline the presentation process by breaking down complex concepts so that the average reader may make sense of the data. Concept mapping is a technique that can demonstrate how people visualise relationships between various concepts (Lanzing, 1996). Concept maps are a relative of cognitive maps in psychology, because they provide a visual representation of how people localise their knowledge and construct ties between various concepts. Generally, concept maps have been deployed in social sciences and qualitative research-based studies. Concept maps consist of labelled concepts, linking said concepts, hierarchies, and other visual representations that might help researchers better understand the participants' minds and thought processes regarding the

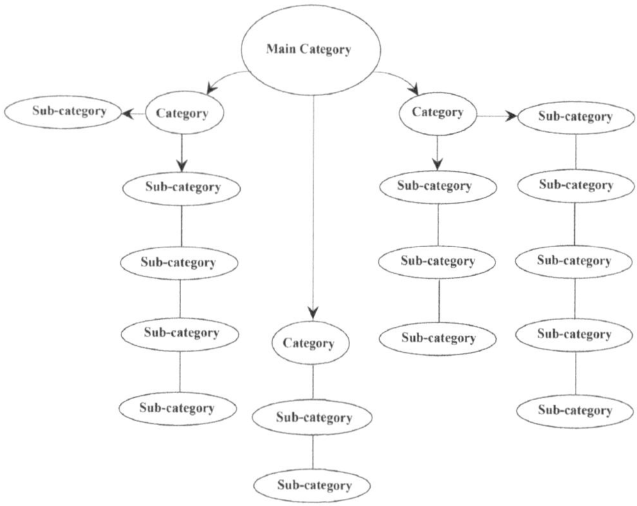

Figure 4.3 Example of Mind Map #3

selected links and relationships between concepts. The visual representations might include overlapping circles, arrows from one concept to another, word links or line links from concept to concept (see Figure 4.3).

Conclusion

Mind mapping can be a useful tool for researchers when designing a qualitative research study. When the investigation requires a look behind the curtain of how respondents view certain concepts and form the relationships between them, mind maps are the often the appropriate methodology to use. This chapter aimed to discuss how people evaluate, analyse and compartmentalise information and how this may be revealed by researchers through the use of mind mapping techniques. We discussed the history of mind maps, the step-by-step process to conducting studies using both a mind map and a semantic. Finally, we pondered upon the pros and cons of these techniques. Mind maps are incredibly valuable if properly understood and deployed correctly within a research study. That being said, we shall conclude with three mind mapping tips.

Use Differentiators

When constructing a mind map of your own, it is helpful to use colour, a variety of line styles (darker lines, dotted lines, etc.) various shapes to represent importance or other aspects of the study that need to be differentiated clearly on the map. Not only will this help make the map easier for the researcher to digest information, but it will also give the participant the opportunity to be creative and express their full self. Just make sure to define the visual language before starting, so that the results and insights will remain clear, concise and unambiguous to the participant and to later readers of the study's results.

Simplicity Is Key

Nowadays, there are a lot of mind mapping software tools available on the web. What we have provided in our examples are the basic paper and pen applications as software may not be available. Furthermore, being familiar with manual applications of mapping procedures helps you to better understand and use digital renditions of the same tests. If you know how to use the digital versions then these can be great resources. However, learning to use the software may be considered by some researchers, especially those using a new mapping technique in a small-scale study, that adopting new software with its learning curve to mastery before application can overcomplicate the process. Typically, the concept that the researcher is investigating and trying to understand by using a mapping approach, is quite intricate and complex. Therefore, the simpler the mind map is, the easier it will be to use, to generate meaningful insights and comprehend and use these.

Physical Maps

Using pen and paper may seem primitive with all the technology we have these days. However, constructing a mind map by hand can be beneficial because it enhances memory through actively involving participants in the construction process. Preference and situation are important here. For instance, if the study is being conducted globally or your team is located elsewhere, employing a digital or remote mind map may be the best solution. The same may be true if your sample participants are young and used to digital platforms. In this situation, they may feel more at home in the digital world whilst older people may be happier in an analogue world.

Kindly note that these aren't the mind mapping standards but instead are our personal tips to constructing a powerful and effective mind map.

References

Buzan, T. (2018). *Mind Map Mastery: The Complete Guide to Learning and Using the Most Powerful Thinking Tool in the Universe*. London: Watkins Publishing.

Casey, E.S. (2002). *Representing Place: Landscape Painting and Maps*. Minneapolis: University of Minnesota Press.

Cassirer, E. (1946). *Language and Myth*. New York: Harper & Brothers.

Downs, R.M., and Stea, D. (2018). *Image and Environment: Cognitive Mapping and Spatial Behavior*. New York: Routledge.

Glaser, B.G. (1965). The Constant Comment Method of Qualitative Analysis. *Social Problems*, 12(4), 436–445.

Habermas, J. (1976). *Communication and the Evolution of Society*. London: Polity Press.

Hathaway, A.D., and Atkinson, M. (2003). Active Interview Tactics in Research on Public Deviants: Exploring the Two-cop Personas. *Field Methods*, 15, 161–185.

Husserl, E. (1970). *The Crisis of European Sciences and Transcendental Phenomenology*. D. Carr (trans.). Evanston: Northwestern University Press.

Korzybski, A. (1933). *Science and Sanity: An Introduction to Non-Aristotelian Systems and General Semantics*. Fort Worth: Institute of General Semantics.

Lanzing, J.W.A. (1996). *Everything You Always Wanted to Know About . . . Concept Mapping*. Retrieved October 8, 2005. http://users.edte.utwente.nl/lanzing/cm_home.htm

MacEachren, A.M. (2004). *How Maps Work: Representation, Visualization, and Design*. New York: The Guilford Press.

Tattersall, C., Watts, A., and Vernon, S. (2007). Mind Mapping as a Tool in Qualitative Research. *Nursing Times*, 103(26), 32–33.

5 Developing Your Own Unique Projective or Sort Approach

Chapter Summary

In this chapter, we consider the fact that there is no single way to design and conduct research that is based upon or includes a projective technique. We note how these approaches are very frequently used within a focus group setting and that they may be used to accompany and compliment other forms of data gathering. We then take the reader through step-by-step approach to how a novel projective technique may be developed in order to meet the specific needs of a particular piece of research. We consider how a researcher may wish to attempt to understand how participants think and feel about candidates in a political election. We suggest that the researcher may decide to develop a novel form of an empty balloon procedure to delve into participants' understanding of the candidates. We offer three variations of this basic design and suggest the type of information that the three variations may produce and the insight they may provide. An example is also provided of a novel technique that we have designed specifically as an illustration for this book, which employs the gathering of digital visual and sound files, in order to illustrate a participant's disposition towards an event or object.

Introduction

Up to this point in the book, we have reviewed projective tests of many types. One thing that all of these tests have had in

DOI: 10.4324/9781003285892-6

common is that they have been previously used in human-subject research. That is to say, they exist and the researchers have been able to choose one or more of these tests for their research. There are, however, times when existing techniques may not be a perfect fit for the research in hand. Furthermore, projective approaches are very frequently used within focus group research and may need to be adapted to the specific context of the group or to accompany and compliment the other forms of data gathering that are being used. In this situation, the researcher may choose to adapt an existing projective test or to develop a totally new projective procedure. If the researcher chooses to develop a new projective technique, the first thing that they must be aware of is that there is no single way to design and conduct human subject research that is based upon or includes a projective technique within its design.[1]

In this chapter, we take the reader through how a novel projective technique may be developed in order to meet the specific needs of a particular piece of research and provide an example of how to design a novel projective tool.

Introduction to Developing Projective or Sort Approaches

It is worthy of note that developing a projective method for your research project may not be an overly complicated, time consuming or expensive enterprise. Unlike psychometric tests and questionnaires, projective methods are attempting to gather information that is in essence a snapshot that is valid at the time and place of its being collected but typically not beyond. As such a newly developed projective technique does not have to go through stringent and lengthy piloting and the data gathered does not have to be analysed from different samples in order to develop an understanding of how responses from the new tool are related to the population to which the researcher is attempting to compare the information he or she has collected in a new study. This does not mean that projective techniques can be thrown together. The opposite is true and much careful preparation much be gone through. A new projective test will also need to be piloted in order to see if it works, if it is understood by participants, and to see if it produces the type of information the researcher was hoping for. It should be remembered that as researchers, we may understand the area of research in a totally different manner to the participants in our study. We know that a projective test will produce responses that answer the questions we are asking in the way we are asking them, and that respondents are understanding the procedure as we wished them to, only after we have asked a few respondents to offer their responses to our projective procedure: only after such piloting can we be sure the participants understand what is being asked of them in the way intended.

Below we will suggest ways of developing a projective approach. However, before we provide an illustration as to how to develop a new projective approach, we will start by considering the use of projectives in focus groups and other settings as this will demonstrate how the need to develop a novel technique may arise and be met.

Projective Approaches and Focus Group Research

We are using the example of using projective techniques with focus groups as this practice is both common place and is also illustrative of how a projective technique may be developed to address the specific research questions and context of the group (see Hackett, 2015; Hackett et al., 2016). The reason for projective techniques frequently being included within focus groups is that these groups have the aim of generating new ideas and novel solutions to problems. Similarly, projective techniques may have these aims and represent a way of delving deeper into the thoughts and emotions of the participants, either as individuals or as a group (Cui et al., 2018; Pich and Dean, 2015; Varela et al., 2022).

Other Forms of Data Gathering

Focus groups are not the only other types of research studies that projectives may be used in tandem with. For example, in-depth interviews are a means for exploring an individual's knowledge of or beliefs about an event or phenomenon that allows the researcher to probe and ask follow-up questions to achieve clarity and well-constructed answers to the research questions. When used in this context, projective techniques may be used before, during or after the interview. When used before the interview, the researcher has the intention of using the technique to reveal information about the topic of the interview that may be used to generate or modify the subsequent interview. When used during the interview, the researcher will ask questions and will then use the projective procedure to elaborate upon these prior to returning to the interview to further explore these issues. When employed after a focus group, the projective technique may be used to explore the information that has been revealed during the in-depth interview.

An Example of the Development of a Novel Projective Research Approach

When designing your own unique projective technique, the first thing the researcher must do is to refamiliarise themselves with the overall research topic and the questions and aims of the research project. They then need to ask themselves why they are considering including a projective approach in their project. The answer to this question may take the form of an answer such as: to try to tap into the unconscious knowledge or motives behind respondents' verbal responses; to enable the participant's ability to physically manipulate stimulus material and to thus reveal the way they feel parts of the research topic relate to each other, or; to reveal the way the respondent thinks about a particular concept or topic and the underlying associations that a respondent has in regard to this topic. There are many other answers to the question of why a researcher is considering using a projective technique in their research, but the three examples will serve to illustrate how knowing why the projective approach is being considered a good starting point when attempting to design a new projective technique. If we look at each of the three example answers to why the projective technique is being seen as an option, then we will be able to see that specific projective approaches appear to be most appropriate for each answer.

For example, with the first answer, *to try to tap into the unconscious knowledge, feelings or motives behind respondents' verbal responses*, an approach is needed that presents the participants' verbal responses back to them in a way that requires and enables them to provide further responses that are not verbal. An example of a projective technique that is suitable in this situation may be a sort procedure in which the main aspects of the respondents' verbal replies are presented to the respondent on cards and they are asked to perform a triadic sort of the cards, which is a procedure in which a respondent identifies why two of the terms on the cards are similar to each other but from the third term. A triadic sorting procedure would also facilitate further discussion around why the respondent has sorted the cards in the way they have. In the second example, *to enable the participant's ability to physically manipulate stimulus material and to thus reveal the way they feel parts of the research topic relate to each other*, cards with words or phrases could again be used, or physical objects could be the items that are sorted, and participants could be asked to sort the items in any way they wish in order to reflect how they felt the items related to each other. As with the first example, this sorting would then form the basis of further discussions and explorations as to why the items were sorted as they were. In the final example, *to reveal the way the respondent thinks about a particular social concept or topic and the underlying associations that a respondent has in regard to this topic*, an empty balloon technique may be used that requires the participant to supply words or phrases to fill empty speech balloons for pairs of stick figures which are presented to them and where the researchers specify the situations in which the stick figures are supposed to be located. The figures may be the participant or other specified individuals or they may be un-named. As with other projective techniques, discussion and exploration of responses will follow.

If we take the last of these examples, the speech balloons used *to reveal the way the respondent thinks about a particular social concept or topic and the underlying associations that a respondent uses in regard to this topic*, we can provide details about how a researcher could design a projective procedure that is unique to this study. A few further details of this imaginary study are first needed in order to provide a contextual background to the projective we will design. If we imagine we are interested in how people will vote in an upcoming election for a local politician, we would start by asking ourselves what it is we are hoping to achieve or reveal through using a projective technique in this situation. Our answer may be that we wish to reveal and to understand an individual's thoughts and feelings towards each of the prospective candidates. Of course, we could simply ask participants in our study questions such as: "what do you think about candidate X?"; "what do you think about candidate Y?"; and "what do you think about candidate Z?" Such questions would be likely to reveal what the respondents consciously think about the candidates and the rationalisation the participants use to filter and justify their thoughts and their verbal reports of these. Such rational reports may provide information for our study but are likely to miss many of the more subconscious beliefs and feelings a respondent holds.

With this aim in mind, we could design empty balloon procedures as follows.

First, we could produce some printed or digital images of two stick people located against the background of an image that represented a participant's front door (there would be text under the image explaining that the door is your front door and the researcher would also explain this). Also under the one stick figure would be the word "you" and under the second figure, the name of one of the candidates. The researcher would explain to the participant that the image depicts the scenario in which the candidate comes to you front door and introduces themselves to you. You are asked to first fill in your empty speech bubble with the most pressing question you would like to ask the candidate and then to fill in the candidate's bubble with the response you believe they may make. By doing this, it may be possible to demonstrate respondent's most pressing uncertainty they have with a candidate (as revealed in the candid question) and also to demonstrate their feelings about the candidate through the response they imagine the candidate will give to them.

A second empty balloon procedure we could use may be designed to reveal how the candidates are differentiated by the respondent. In order to address this aim, we could produce a different empty balloon procedure in which we have as many stick figures as there are candidates in the election. Each stick figure would have a candidate's name printed below it and the respondent would be asked to fill in an empty speech bubble for each candidate. In this example, the background would be blank and the researcher would explain that the respondent should imagine all of the candidates met in a neutral setting and that each candidate was able to ask a single question of another single candidate. The respondent would then be asked to fill in the response bubble for each candidate specifying to whom the question is aimed and what the question is. This procedure may reveal how the respondent sees each candidate's orientation towards the important questions that determine the upcoming election. It may also show how the respondent thinks that each candidate has or does not have answers that differentiate each of the candidates.

A third projective approach could be incorporated into the research project which uses the same image with multiple stick figures representing each of the candidates and each figure having an empty speech bubble above their heads. This time the researcher would explain to the respondent that the candidates are all prepared to answer the same single question from them. The respondent is then asked what their question would be and the respondent is also asked to fill in a response for each candidate. By using this approach, it would be the hope of the researcher to identify the single biggest issue for the candidate. They may also be able to assess the relative strengths and weaknesses of each candidate in terms of this salient issue.

Closing Summary of Adapted Technique

In this chapter, we have emphasised the fact that projective techniques are varied and very adaptable to the needs of a specific research question and situation. In order to support this assertion, we provided an example of an imaginary research study into political candidates and how voter understanding of the candidates may be tapped using projective approaches that used the empty balloon technique. In the above sections, we have aimed to illustrate how a simple projective technique can be conceived of and then designed in a way that would address a specific need in a fictitious research project. The example provided also shows how a projective technique can be adapted so that it incorporates a series of differences in what is essentially the same procedure; in our case, this was the empty bubble approach and we hope that our illustrative example will encourage readers to consider adapting projective approaches in order to specifically meet their own needs.

We also said at the start of the chapter that after we presented the adaptation of an existing projective, we would illustrate how to design a new projective approach. In the next, and final, section of this chapter, we present such an illustration.

A Novel Approach

It should be stressed that whilst the example provided earlier was an adaptation, the widely used empty bubble approach for gathering qualitative data, the researcher does not have to adapt an existing approach and may design a procedure totally from their imagination. For instance, through brainstorming, we have just come up with the idea of creating a more intricate and perhaps a slightly longer longer-term projective approach.

In this imaginary study, the participant is given a specific topic and it is explained that this is the area the research is about. This topic may be about a thing, an event, an object, a product, service and so on, or it may be of a state of being, such as happiness, sorrow and contentment. The participant is then asked to use their phone to record still and moving images and to record any pertinent sounds and/or their own spoken thoughts about the topic they have been given. They would be told that entries should reflect how they are thinking and feeling about the topic they have been given. Participants are instructed to make at least one entry per day for a specified period of time. They are also told that they can however make as many entries as they wish. The time over which the procedure will run would be determined by the researcher, which could be just a couple of days or up to, for example, a week. It would probably be better to stick to a week as the maximum so as to maintain the interest of participants and the quality of their entries.

Once the material has been gathered by the participant, the researcher will have to decide how they wish the respondent to present the material to them. They could, for example, ask them to go through what they have recorded whilst the researcher is present, explain the reasons for the inclusion of the image or sound and provide details around their spoken contributions. An alternate and more complex way of presenting the material would be to ask the participant to assemble a social media page that displays the material they gathered arranged in a way that allows the participant to best convey the message they wish to convey.

Conclusion

We hope that the illustrations of adapting an approach and developing a totally new approach have provided readers with a clear understanding in regard to the design and modification of projective techniques in order to fit the researcher's questions. In the next chapter, we turn our attention to the potential future application of projective approaches in areas and domains in which they have tended not to have been used.

Note

1 There are many good texts that exist which assist with the design of qualitative human subject research to which interested readers

are guided, for example: Bukve, 2019; Blaikie and Priest, 2019; Hackett, 2015; LeCompte and Shensul, 2020; Marshall, 2021.

References

Blaikie, N., and Priest, J. (2019). *Designing Social Research: The Logic of Anticipation*. Cambridge: Polity.

Bukve, O. (2019). *Designing Social Science Research*. Basingstoke: Palgrave Macmillan.

Cui, C., Mrad, M., and Hogg, M. (2018). Brand Addiction: Exploring the Concept and Its Definition Through an Experiential Lens. *Journal of Business Research*, 87, 118–127.

Hackett, P.M.W. (ed.) (2015). *Qualitative Research Methods in Consumer Psychology: Ethnography and Culture*. New York: Routledge Publishers.

Hackett, P.M.W., Schwarzenbach, J.B., and Jurgens, A.M. (2016). *Consumer Psychology: A Study Guide to Qualitative Research Methods*. Leverkusen, Germany: Barbara Budrich Publishers.

LeCompte, M.D., and Shensul, J.J. (2020). *Designing and Conducting Ethnographic Research: An Introduction* (Volume 1) (Ethnographer's Toolkit, 2nd edition, 1), Washington, DC: AltaMira Press.

Marshall, C. (2021). *Designing Qualitative Research*. Thousand Oaks, CA: Sage.

Pich, C., and Dean, D. (2015). Qualitative Projective Techniques in Political Brand Image Research From the Perspective of Young Adults. *Qualitative Market Research*, 18(1), 115–144.

Varela, P., Arvisenet, G., Gonera, A., Myhrer, K., Fifi, V., and Valentin, D. (2022). Meat Replacer? No Thanks! The Clash Between Naturalness and Processing: An Explorative Study of the Perception of Plant-Based Foods. *Appetite*, 169, 105793.

6 The Future of Projective Technique Research

Chapter Summary

There are very many spheres of enquiry within which projective technique research has the possibility or indeed the likelihood of being employed in the future. This obviously includes the areas in which it has already been applied. It is also likely that when conducting research in areas that have already been investigated, the manner in which these techniques have been used will likely be similar to earlier investigations. However, in this chapter, we will make attempts to predict the more novel ways and situations in which projective techniques will be used in the future, and we will consider unusual approaches and domains of usage. We concentrate upon three such neoteric techniques and domains: using projective techniques in critical social research and with individuals from minority communities; using projective techniques in conjunction with the declarative mapping approach; and using projective techniques in online contexts.

Introduction

In this chapter, we consider the future for research that employs a projective or mapping approach in its design. There are always difficulties whenever we attempt to predict the future in any context, and this is certainly the case in the rapidly changing world of social science and humanities research

DOI: 10.4324/9781003285892-7

methods. However, we will do our best to gaze into our crystal ball and to make suggestions as to some of the ways in which projective and sorting-based approaches may usefully evolve or be adapted in the future. We will consider several areas within which we believe these approaches may be particularly useful. We will, for example, consider the application of projective techniques in critical social research, social justice and inclusion research. We will also consider using projective techniques within the declarative mapping sentence and approach to social research, and we will discuss a little upon how human subject research has recently moved to using digitally based methods.[1]

Critical Social Research

In this section, it is important to first seat out what we mean by "critical social research". Critical social research is a phrase that is employed to describe research that adopts approaches to social enquiry that make efforts to delve underneath initial or surface meanings or appearances. This form of research attempts to reveal less apparent meanings within the social world by becoming involved with the subject matter of their research in a critical manner and not simply accepting the apparent. According to Harvey (1990, 1993, 2012, 2022a, 2022b, 2022c, 2022d), critical social research within the social sciences can trace its roots back to the writing of Karl Marx (Marx, [1887] 1977) and has since been developed by Marxists, feminists, anti-racists, structuralists, film theorists, postcolonialists and others. Harvey (2022a) differentiates critical social research from other methods as critical social research, he says, is not in the business of trying to form causal or pseudo-causal links between concepts in the research domain we are investigating. Neither is it, Harvey says, attempting to establish elaborate theoretically complex systems of beliefs. Harvey contrasts critical social research with other research traditions in the social sciences, such as positivism and phenomenology, where he notes that positivism's main concern is with establishing causal relationships and explanations and where phenomenology sets itself the task of interpreting the meaning of social processes and actions.

Many approaches to research are being re-imagined within a critical framework. For example, recently Michelle Fine and Maria Elene Torre (2021) have focused upon participatory action research within a critical context. Similarly, Heide Levitt (2021) has considered critical-constructivist grounded theory research. These authors, as with all who consider a critical perspective within their research, are concerned with the real-world, everyday setting of their research. For example, critical participatory action research, "focusses intentionally on questions of power and injustice, intersectionality and action" (Fine and Torre, 2021). Critical social research is more than adopting an approach to research that attempts to achieve a sensitivity to the power issues and how these both impact, and are part of, the research context; the researcher who adopts a critical approach is typically motivated by wishing to make a positive difference in regard to social justice. To quote Fine and Torre (2021), "We view critical research as one more resource in, by and for movement for justice".

There are many research orientations that have been employed, critically or otherwise, in enquiries into social justice. In this section, we will consider how the approaches to research we have presented in this book have the potential to make a meaningful impact in the area of social justice research. More specifically, what we are meaning by social justice in the context of using projective techniques is the potential these approaches have to be more inclusive of communities that fall outside of those from the affluent global north.

We believe that projective approaches are potentially of value in social justice research as projective techniques require minimal verbal input from the administrator, and as these approaches are attempting to delve deeply into the motives of respondents, projectives are techniques that hold the potential to reveal the beliefs, attitudes and values of individuals in a way that empowers the participant. This suggests that projective and sort-based methods are sensitive research approaches in the specific context of facilitating the voice of individuals from marginalised communities.

Projective Methods in Critical Research

The main characteristic of all projective methods is approaches reliance upon a substantial input from the participant, and at the same time, the approach requires a lesser degree of interjection by the researcher. Furthermore, the researcher using projective methods attempts to make the input they offer during the research process as non-directive as possible. This clearly places the participant in the driving seat in regard to their being able to both guide the research and provide responses in the ways they wish to. This is of particular value in critical research as critical research, by its very nature and definition, may be seen to be pushing back against hegemonic forces and opinions held by institutions of the establishment, of which the researcher may be a part. Of course, projective techniques are still subject to the participant being part of the institutions of the dominant social institutions and these memberships influencing the participant's responses. However, these and other bias and influences are part of the research process that critical research attempt to mitigate against. Notwithstanding such potential issues, we believe that projective techniques offer a viable way of maximising the strength of the voice of participants in critical social justice research.

Projective Methods as Sensitive Research Approaches

When the research that is being conducted is into an area that the participant may find sensitive, projective techniques may be appropriate methods for gathering information. The reasons for the applicability of projective approaches are many-fold, and include such reasons as that using a projective technique may allow sensitive topics to be presented in a less threatening and in a less direct manner. Projective approaches, by allowing respondents to provide the response they wish in a way they decide, may be less intrusive, or may initiate less in the way of a guarded response, than other more direct forms of enquiry. Additionally, and as with any projective technique research, the approach provides an excellent window that the researcher may open to allow in-depth discussions about the topic that is being addressed. The fact that the participant has guided the research may also mean that they will be more open to respond honestly and openly to follow-up probes from the researcher.

Projective Methods With Individuals in Marginalised Communities

In any social science research project, researchers may, and indeed are likely to be drawn from very different sections of the community to those from which the respondents they are working with come from. The values that researchers hold, and even the language they use, may also differ significantly to the language and values of the people they are working with. Projective techniques, whilst they will not completely remove these potential difficulties and biases, may help to work against these potential problems as they require lesser degrees of input from the researcher. Furthermore, researchers may be able to more readily understand the responses given by a participant as the participant wished them to owing to the fact that the response is often not linguistic and the respondent may elaborate upon a response rather than simply giving a verbal reply to a question.

So far in this chapter, we have briefly considered the nature of critical social research and the use of projective techniques within specific areas: marginalised communities and when addressing sensitive research questions. In the paragraphs that follow, we will offer some suggestions regarding the use of projectives and mapping approaches when using a declarative mapping approach as your research design.

The Declarative Mapping Approach

The projective and mapping/sorting techniques we have presented in this book have been considered by us stand-alone research methods. However, all research exists within a context, which may be made up of events and circumstances outside of the research itself or other forms of information gathering within a broader research project. Furthermore, many research projects may be complex in nature and attempt to simultaneously consider multiple aspects of a research situation in their efforts to provide the answers to the questions and aims of its research. Such intricate research is often difficult to effective design, conduct, manage and analyse. In this section, we present the declarative mapping approach to conducting complex types of research (Hackett, 2014, 2020, 2021a, 2021b; Hackett and Fisher, 2019; Hackett and Gordley-Smith, 2022a, 2022b; Hackett and Li, 2022; Hackett and Lustig, 2021; Schwarzenbach and Hackett, 2015).[2] The declarative mapping sentence provides a flexible and reflexive framework within which to both design and analyse complex research enquiries, and we propose that declarative mapping sentences may be used with profit in research designs that include complex projective techniques that may consider the context.

The declarative mapping approach to social science research is based within a descriptive philosophical framework that is developed to become a method of explication and elucidation. The qualitative researcher who employs the use of the declarative mapping approach uses its major tool, the declarative mapping sentence, within which they specify the major subparts of a domain that is the focus of research interest. The declarative mapping sentence also posits the relationships between these major sub-components (Hackett, 2014, 2018, 2020, 2021a, 2021b) that have been applied to study numerous social science and humanities research topics from a variety of disciplines (e.g. Hackett, 2016, 2017, 2019; Hackett and Gordley-Smith, 2022a, 2022b; Hackett and Li, 2022; Schwarzenbach and Hackett, 2015).

Declarative mapping sentences are explicit statements of the part-to-whole and part-to-part relationships of the major sub-components of the domain that the researcher is studying. By developing and stating a declarative mapping sentence for a domain, the researcher attempts to bring clarity and appreciation of the research in a way that makes clear how the sub-aspects of the research domain interact. Moreover, the declarative mapping sentence provides a flexible and adaptable framework for designing research instruments to examine the domain that is mapped. The declarative mapping sentence is written in normal language and constitutes a framework or structure that the researcher tailors to a specific research situation in response to the findings the researcher reveals as they conduct their research project. Declarative mapping sentence research, by its explicit nature, constitutes a transparent and rigorous approach to conducting research which allows the results from different declarative mapping approach projects to be compared. Because the domain that is being studied is overtly stated within the sentence, the imposition of researcher's interpretative biases is reduced.

We have very briefly described the declarative mapping approach to research, which we believe would work well together with projective research techniques. The reason for our holding belief is that the Declarative mapping approach provides a framework for designing research and interpreting the results of a research project that has been designed within the sentence's rubric. Projective approaches, on the other hand, offer a set of tools for gathering information in a respondent-driven approach that allows for the respondent's voice to be at the forefront of the research. We believe that bringing the two approaches together would have specific benefits and an illustrative example will help the reader to appreciate better this synthesis of approaches.

If we take the example of our wishing to gauge a sample of respondents' attitudes and beliefs towards their experiences in a nature reserve, we may decide that we want to understand users' responses to different aspects of the nature reserve through a triadic sort approach. Having made the decision to use this projective research method, we are faced with determining the content of the materials that we wish our respondents to sort. If a declarative mapping approach was adopted and used in conjunction with the triadic sort, then the first task would be for the researcher to review literature, publications and other reports and writing about the reserve, along with having initial, open, conversations with individuals about their experiences in the location. This review that they have undertaken would, it is hoped, identify the major aspects of nature reserve experience, or any other themes that participants feel to be important. For example, after engaging in such an initial review, the researcher may identify the following themes: services at the reserve; social aspects of the reserve; directness of contact with aspects of the reserve and pertinence to the user in order to achieve their aims whilst at the reserve. Having identified these major themes of reserve experience, the researchers would then write these as a declarative mapping sentence in which the themes are linked with ordinary language phrases in a manner which suggest the associations between themes. An example of this declarative mapping sentence would be:

> A person experiences the nature reserve in terms of the *services at the reserve* and the *social aspects of the reserve* with which they have *different levels of directness of contact* with aspects of the reserve as these are *pertinent to user in order to achieve their aims* whilst at the reserve.
>
> The major themes, that are known as facets, are highlighted in italic script.

The researchers would then need to pilot-test these issues to see if they had managed to correctly identify these and that they made sense to respondents (we will not go into details of piloting as this is not what we are concentrating upon in this example). Having identified the major aspects of the research domain, which are termed "facets", and joined these together in a sentence that suggests how the facets are inter-related, the next task is to identify the sub-aspects of each facet. The sub-aspects of each facet are how an individual actually relates to the reserve in terms of the facet. For example, with the directness of content with the nature reserve's facilities, the sub-components (which are called elements) may be *direct* and *indirect*. The elements of the services at the reserve facet may have elements of *bird/nature watching*, *refreshments*, *car parking* and *reserve information*. Thus, the declarative mapping sentence may be written as:

A person experiences the nature reserve in terms of the
(services at the reserve) – 1

- bird/nature watching – a
- refreshments – b
- car parking – c
- reserve information – d
- toilets – e

services, and
(social aspects of the reserve) – 2

- access to reserve staff – a
- chances to be alone – b
- meeting/discussion areas – c
- signage – d

with which they have
(different levels of directness of contact) – 3

- direct – a
- indirect – b

Levels of contact, with aspects of the reserve as these are
(pertinent to user in order to achieve their aims) – 4

- more pertinent – a
- less pertinent – b

to them achieving their aims whilst at the reserve.

(N.B.: The letters after the facet names and the letters after the elements are included so as the facets and elements may be identified in the writing in this book and are not a necessary part of a declarative mapping sentence.)

The declarative mapping sentence that the researcher has assembled forms a template for designing any projective procedures so that it addresses their research questions. In our example, by selecting elements from each facet and combining this with elements from other facets, a project technique may be tailored to comprehensively focus upon user experience at the nature reserve. To take a more specific example, if we decided to use a triadic sort technique, we could design this to have the features of the reserve's experience as these are specific in the declarative mapping sentence, written on cards for the respondents to differentiate. Each word or phrase on a card would be typified by a profile that is made up of a combination of one element from each facet. For example, if our sample was of bird watchers, the phrase "visibility from bird hides" could be printed on one card. Given the sample in our study, this phrase would have a profile of: 1a, 2b, 3b, 4a. Another example phrase could be "food in the reserve cafe" which would have a profile of: 1b, 2c, 3b, 4b. The profiles are suggestive and only after the research has been conducted and analysed can the researcher have confidence in the profiles. For instance, we have suggested in the second example profile that the cafe's facilities are of less pertinence to birdwatchers but this may not be the case for every particular sample of respondents.

As well as being a framework for designing comprehensive research, the declarative mapping sentence forms a structure around which the information that comes out of procedures such as triadic sorts may be analysed. Finally, it is also a framework for structuring the write-up of the results from a research project.

The Move to Digital Research Methods

Over the course of the past several decades, much of the world has moved online. The transference of research so as to incorporate online methods and data is very much the case when we look at research into human behaviour and experience (e.g. see Przybylski, 2020; Tensen, 2019; Hackett and

Hayre, 2021a). The advent of the global COVID-19 pandemic accelerated the move of research to be conducted within a digital context as the disease made face-to-face contact problematic or prohibited in many contexts: This was especially within social science and humanities research. As a consequence, during the pandemic, the move to more online forms of research was hastened and many new digital/online techniques were developed and existing methodologies have been rapidly adapted for use within a digital context (e.g. see Accelerant Research, 2022). Online methods came to the fore at this time and even those who had previously resisted using a digitally based research approach found themselves with little choice but to adopt one. In the section below we consider the move online for projective approaches and what the future may hold.

For the qualitative researcher, projective techniques lend themselves extremely well to an online setting and researchers have hastily and vigorously adopted them. Much of the qualitative research that is conducted online may employ software-based interactive components and online projective techniques may be designed and administered in order to facilitate many of the aims of the qualitative researcher. For example, online projective tests may be (to some extent) standardised in that they can be designed and then administered in a consistent manner. Online projective approaches may also be semi-structured and this feature enables for greater and easier comparability between different applications. Moreover, online applications of projective procedures may allow a researcher to take advantage of the positive aspects of projective approaches in a potentially cheaper format and one which does not require participants and researchers to have to travel and be physically present together during the delivery of the procedure. When giving tests online, it is usually a good idea for the researcher to provide ample information on how to complete a test as the researcher is not physically present, which would allow interaction with the participant about any uncertainties they may encounter. However, when using these techniques online, it must be remembered not to provide participants with too much information in regard to how they complete the projective procedure as you are attempting to gain spontaneous insight from the respondent. It should also be remembered that many offline projective applications can be readily and highly effectively adapted to the online context.

Examples of projective approaches that have been adapted to a digital format include many of the text-based methods. Text-based approaches, such as sentence or story completion, obituaries and billboards, are obvious candidates for digital and online presentation as they simply involve the participant in writing a response onto a computer. Sort approaches have also been successfully adapted to the digital realm. We are all probably familiar with online and digital card games, and the sort procedures are close analogies to these and readily can include text and images formatted sort procedures. Most other types of projectives approaches, such as creating mental maps, can successfully be moved to a digital and online form of usage. The greatest challenge perhaps are techniques that require the manipulation of physical objects in sorts or other techniques.

As we write, during the latter months of 2022, the COVID-19 pandemic is less severe than it was, and most restrictions on activities have been removed in most countries. Notwithstanding this return to some form of normality, digitally-based research has taken a huge step forward due to the pandemic, and due to the ease, efficiency and cost-saving characteristics associated with locating research in this realm, it is likely to become more and more common-place. Moreover, much of the world has interacted online and digitally for three plus decades and increasing proportions of the world are extremely comfortable, and even expect, to conduct a lot of their lives online, including the completion of research projects in which they participate.

Conclusion

In this chapter, we have gazed into our crystal-ball and made prophecies in regard to how we believe that projective techniques may be used in the future. We may well be incorrect in our predictions, as the future is typically extremely difficult to foretell. However, in spite of such challenges, we hope we have indicated that projective techniques are evolving procedures that we believe will be employed by a wide range of professional and academic researchers who wish to gather information to provide insight into how people think, feel and behave.

Notes

1 Interesting examples of analogous research can be found in: Hackett and Hayre, 2021a, 2021b; Hackett and Schwarzenbach, 2021.
2 The declarative mapping sentence was developed out of the traditional mapping sentence (see Levy, 1976, 1985, 1990) which is a major component of the facet theory approach to social science research (see Borg and Shye, 1995; Canter, 1985; Dancer, 1990; Guttman, 1954; Hackett, 2014; Shye et al., 1994; Tziner, 1987).

References

Accelerant Research (2022). *Online Projective Exercises Breathing New Life Into Classic Qualitative Research Techniques*. The Greenbook Directory. Retrieved from www.greenbook.org/marketing-research/online-projective-exercises-37076

Borg, I., and Shye, S. (1995). *Facet Theory: Form and Content, Advanced Quantitative Techniques in the Social Sciences Series*. Thousand Oaks, CA: Sage Publishers.

Canter, D. (ed.) (1985). *Facet Theory: Approaches to Social Research*. New York: Springer Verlag.

Dancer, S. (1990). Introduction to Facet Theory and Its Application. *Applied Psychology: An International Review*, 39, 365–377. *Google Scholar*

Fine, M., and Torre, M.E. (2021). *Essentials of Critical Participatory Action Research (Essentials of Qualitative Methods)*. Washington, DC: American Psychological Association

Guttman, L. (1954). A New Approach to Factor Analysis: The Radex. In P. Lazarsfield (ed.), *Mathematical Thinking in the Social Science* (pp. 258–348). New York: Columbia University Press. *Google Scholar*

Hackett, P.M.W. (2014). *Facet Theory and the Mapping Sentence: Evolving Philosophy, Use and Application*. Basingstoke: Palgrave Macmillan.

Hackett, P.M.W. (2016). *Psychology and Philosophy of Abstract Art: Neuro-aesthetics, Perception and Comprehension*. Basingstoke: Palgrave Macmillan.

Hackett, P.M.W. (2017). *The Perceptual Structure of Three-Dimensional Art, Springer Briefs in Philosophy*. New York: Springer.

Hackett, P.M.W. (ed.) (2018). *Mereologies, Ontologies and Facets: The Categorial Structure of Reality*. Lanham, MD: Lexington Books.

Hackett, P.M.W. (2019). *The Complexity of Bird Behaviour: A Facet Theory Approach*. Cham, CH: Springer.

Hackett, P.M.W. (2020). *Declarative Mapping Sentences in Qualitative Research: Theoretical, Linguistic, and Applied Usages*. London: Routledge.

Hackett, P.M.W. (2021a). Some Questions Regarding the Properties of the Declarative Mapping Sentence Within Facet Theory Ontology. *Academia Letters*. https://doi.org/10.20935/AL3577

Hackett, P.M.W. (2021b). *Facet Theory and the Mapping Sentence: Evolving Philosophy, Use and Declarative Applications* (2nd, revised and enlarged edition). Basingstoke: Palgrave Macmillan.

Hackett, P.M.W., and Fisher, Y. (eds.) (2019). *Advances in Facet Theory Research: Developments in Theory and Application and Competing Approaches*. Lausanne, Switzerland: Frontiers Media SA.

Hackett, P.M.W., and Gordley-Smith, A. (2022a). Describing Racist or Racialised Actions Using the Declarative Mapping Sentence Method. *Frontiers in Sociology: Race and Ethnicity*. https://doi.org/10.3389/fsoc.2022.779452

Hackett, P.M.W., and Gordley-Smith, A. (2022b). An Exploration and Exposition of Paulin Hountondji's Philosophy through a Declarative Mapping Sentence Approach. *Academia Letters*, Article 5096. https://doi.org/10.20935/AL5096. ISSN: 2771-9359

Hackett, P.M.W., and Hayre, C.M. (2021a). Conclusions and the Future of Ethnography in Health-related Research. In P.M.W. Hackett and C.M. Hayre (eds.), *Handbook of Ethnography in Healthcare Research* (pp. 513–521). London: Routledge.

Hackett, P.M.W., and Hayre, C.M. (2021b). Coronavirus and COVID-19: Qualitative Healthcare Research During and After the Pandemic. In P.M.W. Hackett and C.M. Hayre (eds.), *Handbook of Ethnography in Healthcare Research* (pp. 503–512). London: Routledge.

Hackett, P.M.W., and Li, C. (2022). *Using Declarative Mapping Sentences in Psychological Research: Applying Facet Theory in Multi-Componential Critical Analyses of Female Representation in Science Fiction Film and Television*. London: Routledge.

Hackett, P.M.W., and Lustig, K. (2021). *An Introduction to Using Mapping Sentences*. Basingstoke: Palgrave Macmillan.

Hackett, P.M.W., and Schwarzenbach, J.B. (2021). Black Lives Matter; Birdwatching in Central Park and the Murder of George Floyd. In P.M.W. Hackett and C.M. Hayre (eds.), *Handbook of Ethnography in Healthcare Research* (pp. 513–521). London: Routledge.

Harvey, L. (2022a). Critical Social Research: Re-examining Quality. *Quality in Higher Education*, 28(2), 145–152. https://doi.org/10.1080/13538322.2022.2037762

Harvey, L. ([1990] 2022b). *Critical Social Research*. London: Unwin Hyman. Retrieved from www.qualityresearchinternational.com/csr/

Harvey, L. ([2012] 2022c). Critical Social Research. In *Researching the Real World*. Retrieved from http://www.qualityresearchinternational.com/methodology/RRW2pt4Critical.php

Harvey, L. ([2012] 2022d). Critical Social Research. In *Social Research Glossary*. Quality Research International. Retrieved from www.qualityresearchinternational.com/socialresearch/csr.htm

Harvey, L., and MacDonald, M. (1993). *Doing Sociology: A Practical Introduction*. London: Macmillan.

Levitt, H. (2021). *Essentials of Critical-Constructivist Grounded Theory Research (Essentials of Qualitative Methods)*. Washington, DC: American Psychological Association

Levy, S. (1976). Use of the Mapping Sentence for Coordinating Theory and Research: A Cross-cultural Example. *Quality and Quantity*, 10, 117–125. https://doi.org/10.1007/BF00144163

Levy, S. (1985). Lawful Roles of Facets in Social Theories. In D. Canter (ed.), *Facet Theory: Approaches to Social Research*. New York: Springer. *Google Scholar*

Levy, S. (1990). The Mapping Sentence in Cumulative Theory Construction: Well-being as an Example. In J.J. Hox and G. de Yong (eds.), *Operationalization and Research Strategy* (pp. 155–177). Amsterdam: Swets & Zeitlinger. *Google Scholar*

Marx, K. ([1887] 1977). *Capital*. London: Lawrence and Wishart.

Przybylski, L. (2020). *Hybrid Ethnography: Online, Offline, and in Between (Qualitative Research Methods Book 58)*. Thousand Oaks, CA: Sage Publications.

Schwarzenbach, J.B., and Hackett, P.M.W. (2015). *Transatlantic Reflections on the Practice-Based Ph.D. in Fine Art*. New York: Routledge Publishers.

Shye, S., Elizur, D., and Hoffman, M. (1994). *Introduction to Facet Theory: Content Design and Intrinsic Data Analysis in Behavioral Research*. Applied Social Research Methods Series Volume 35. Thousand Oaks, CA: Sage Publishers.

Tensen, B.L. (2019). *Research Strategies for a Digital Age*. Washington, DC: American Psychological Association.

Tziner, A.E. (1987). *The Facet Analytic Approach to Research and Data Processing*. New York: Peter Lang.

7 Conclusion

Chapter Summary

Our final chapter constitutes a very brief conclusion. Within this book, we have presented a wide range of approaches to gathering data that fall under the umbrella of projective techniques. These have, to a very large extent, been approaches that collect qualitative data that are often thought of as subconscious or at least not within the usual arena of awareness for, and easily controllable by, the individuals from whom the data is being gathered. In this final chapter, our writing will aim to simply bring together all the writing that has been included in the preceding six chapters, and we add our concluding remarks. We also briefly outline the intended role of the supplements that follow this final chapter, and these are a glossary of terms and our suggestions for further reading.

Summary of the Book

We started this book by reflecting upon the fact that there are many approaches to conducting research that fall within the disciplines of the social sciences and humanities. We also commented upon our observation that there was a lack of a single book that clearly and exclusively presented the topic of projective technique research approaches and did so in a manner that would form a useful and readily understandable guide for readers who were unfamiliar with such approaches to research.

DOI: 10.4324/9781003285892-8

Throughout this succinct book, we have, as authors, attempted to convey our belief that if you are a scholar or a practitioner from a wide range of disciplines that take human behaviour and experience as their subject matter, projective research approaches are extremely useful and valid ways of gathering research data. We trust that what we have written convinces you that our belief in projective research is well founded (if indeed you needed convincing). It is also our hope that you will be willing to consider using projective approaches in your future research. We further hope that we have conveyed our passion for these approaches and the unique way they are able to represent the participant's perspective within the social science research context.

One of the central aspects of the approaches that we have included in this book, and indeed one of the principal components of the projective approaches we did not have space to include, is the fact that projective techniques may be adapted to a wide and infinitely varied range of research questions, problems and situations. We want to stress that, perhaps uniquely amongst research techniques that are focused upon revealing information about human behaviour and experience, using projective approaches requires the researcher to be creative and imaginative in conceiving, designing and conducting their research.

The illustrative approaches we have included in the book are not supposed to be "set in stone" although many of these have been used for many years in approximately the format that we have detailed in the book. Instead, it is our hope that you will take the essence of each of these examples and consider a projective approach in relation to your own research problems and then make suitable adaptations and produce your own distinctive, idiosyncratic and quirky ways of gathering information.

Furthermore, due to the limitations of space in this slim volume, we have not offered an in-depth consideration of the ways in which projective techniques may be used in conjunction with other approaches for gathering data (although we have briefly considered this to some extent in earlier chapters). However, the ability for projectives to be used as part of a multi-method research design is another feature that recommends these approaches to the researcher as a multiple-method design, by its very nature, may allow a more complete picture to be developed of the research problem that is being addressed. Furthermore, as noted in the preceding chapter, over the past three decades, the world, and specifically the world of research, has moved to the digital arena or online (Hackett and Hayre, 2021). We have not considered this in detail in this book, but many of projective approaches have been adapted, or simply moved to be used on some sort of electronic device. We have not focused upon such digital ways of presenting projective techniques as the basic essence of each technique remains the same regardless of the medium through which it is presented to a participant. We have therefore concentrated on attempting to convey the thinking behind a technique, and we believe that readers will be able to readily use any digital version of an approach without general instruction from us.

Conclusion

We close this book by simply wishing you good luck in using projective techniques in your research, and we also hope you enjoy a sense of satisfaction in your research.

Reference

Hackett, P.M.W., and Hayre, C.M. (2021). Conclusions and the Future of Ethnography in Health-related Research. In P.M.W. Hackett and C.M. Hayre (eds.), *Handbook of Ethnography in Healthcare Research* (pp. 513–521). London: Routledge.

Reading List

Below we provide a list of books and articles that are related to projective techniques, their background philosophy and their applied usage. The list is intended to guide the reader who is interested in broadening their knowledge in regard to projective approaches.

Farzin, M., Fattahi, M., and Khareshi, F. (2020). *Qualitative Marketing Research: Projective Techniques with Picture Drawing.* London: LAP LAMBERT Academic Publishing.

Handler, L., and Thomas, A.D. (2014). *Drawings in assessment and psychotherapy: Research and application.* Routledge/Taylor & Francis Group.

Kolb, B. (2018). *Marketing Research for the Tourism, Hospitality and Events Industries.*

Lambert, M. (2019). *Practical Research Methods in Education: An Early Researcher's Critical Guide.*

Lindsey, G., Elliot, R.M., and Maccorquodale, K. (2012). *Projective Techniques and Cross-Cultural Research.* The Century Psychology Series, Literary Licensing LLC.

Lobinger, K., and Brantner, C. (2020). Picture-sorting Techniques. Card Sorting and Q-sort as Alternative and Complementary Approaches in Visual Social Research. In L. Pauwels and D. Mannay (eds.), *The SAGE Handbook of Visual Research Methods.* The Sage.

Maison, D. (2019). *Qualitative Marketing Research: Understanding Consumer Behaviour.*

McNabb, D.E. (2010). *Research Methods for Political Science: Quantitative and Qualitative Methods.*

Murstein, B.I. (1963). *Theory and Research in Projective Techniques*. Hoboken, NJ: Wiley.

Pauwels, L., and Mannay, D. (2020). *Handbook of Visual Research Methods* (2nd revised and expanded edition) (pp. 309–321). London: Sage.

Soley, L., and Smith, A.L. (2008). *Projective Techniques for Social Science and Business Research*. Palo Alto, CA: The Southshore Press.

Stevens, R.E., Loudon, D.L., Ruddick, M.E., Wrenn, B., and Sherwood, P.K. (2006). *The Marketing Research Guide*.

Tuber, S. (2018). *Using Projective Methods With Children: The Selected Works of Steve Tuber*.

Winston, W., and Percy, L. (1997). *Marketing Research That Pays Off: Case Histories of Marketing Research Leading to Success in the Marketplace*.

Glossary

In this book, we use a series of words, phrases and terms that may have specific or nuanced meanings within the research context. For example, this is especially the case regarding the terms that are used to describe research tools and procedures that we use in this book. Below is a list of the precise meaning of these words and phrases along with further details of the terms and appropriate references. Many of the definitions are taken from Hackett (2019).

Affect (emotion)

Affect is the experience of a feeling or an emotion. These may be simple or complex and may be normal or pathological. Mood and emotions are affective states, which may be positive or negative. Traditionally, affect, conation and cognition are the three components of the mind (American Psychological Association, 2022).

Attitudes

From a psychological perspective, and from the perspective we adopt in this book, attitudes are either thought of a two- or three-componential psychological phenomena. If one takes a three-component perspective on attitudes, the three parts are: affect – how we feel about something; behaviour – what we do about something; and cognitive – what we think about something. The three-part model of attitudes (known as the ABC model) enables the investigation of an individual's orientation towards some thing or event. The relative power of influence of each of these three components will vary from one situation to another. For example, when we decide to skip going to work and have the day on the beach, it is probably affect that is in command. However, when we decide to not buy an expensive car that we like, it is probably cognition winning over affect. Finally, if we buy the same order from a coffee shop every day, it may be behaviour that is in the commanding position. However, it should be noted that over recent decades, the ABC model has been recast as the two-part AC model. This is because research would suggest that behaviour is a separate phenomenon to attitudes. By adopting a two-part understanding, it is possible to investigate the relationships between how people think and feel about something and how they behave in this regard. However, others may argue that this investigation is also possible within the three-part model.

Category

A category is a class or some other form of partitioning or breaking up of things or people on the basis of these having some shared characteristic. In philosophy, categories are usually thought of as "a system of categories is a complete list of highest kinds or genera" (Thomasson, 2018) or as an exhaustive set of classes within which everything may be allocated.

Clause

Traditionally, a clause is said to consist of a subject and predicate. It is a unit of grammatical organisation ordered directly below the sentence. It is the smallest unit in grammar that is able to express complete ideas. A clause may constitute a complete sentence in and of itself or it may be part of a sentence. A main clause is a clause that is able to stand alone and forms a complete sentence as it has a subject and predicate. A subordinate clause is dependent upon a main clause and is a part of this main clause and is usually preceded by a conjunction.

Cognition/Cognitive

Cognition and cognitive refer in some to way to the mental action of thinking and knowing. The American Psychological Association (2018) defines cognition to include awareness and knowing including "perceiving, conceiving, remembering, reasoning, judging, imagining, and problem solving". Furthermore, the American Psychological Association states that traditionally, the components of the mind are cognition, affect and conation.

Cognitive maps

Cognitive maps are a form of psychological depiction or portrayal of a spatial environment. Cognitive maps were first put forward by Tolman (1948). Cognitive maps allow a person to navigate in space, develop spatial coding, fix the location of landmarks and plan spatial routes. In cognitive maps, a person develops a coherent representation of their spatial environment which involves the acquisition and coding storing of spatial details which are later recalled and decoded and used, along with memory inform and guide future action in regard to the locations and attributes of everyday phenomena and events in metaphorical and physical spatial environments. Non-human animal research has suggested that spatial abilities may be located in the hippocampus and related regions. Cognitive maps are representations in neural cells that code place, borders, head direction and borders. In human animals, the hippocampus and entorhinal cortex have been found to maintain spatial codes that are similar to maps, whilst the parahippocampus and retrosplenial cortices allow for the fixing of cognitive maps of environmental landmarks, whilst route planning is associated with the hippocampal and entorhinal spatial regions (Epstein et al., 2017).

Construct

Within this book, we have used the term "construct" in a similar manner to the definition provided by the American Psychological Association (2018). Thus, we take a construct to be an exploratory theoretical model that is based upon empirical research or as a hypothetical model arrived at through inference from empirical data.

Content analysis

Content analysis is the rigorous and systematic analysis of narratives or other forms of non-numerical research. It attempts to identify and describe the underlying or latent content of the texts analysed.

Content domain or content area

It refers to the area that a piece of research addresses (this is clearly specified in the mapping sentence – see below).

Critical approaches to research (critical theory)

Critical research approaches are associated with the evaluative social science research that originated in the neo-Marxist Frankfurt School. Critical approaches intend to be critical rather than simply analysing data that arises out of research. Critical approaches typically take a stance that is counter to orthodox political doctrine with the intention of putting forward and advancing ideas and setting free individuals and groups of people by revealing ideas and practices by individuals and social systems.

Critical race theory

Critical race theory originated in the US and has been defined as a political/intellectual faction and structure that developed within legal analysis. There are many variations amongst critical race theorists as to the precise definitions of race and racism, but there is a series of generally agreed tenets. These tenets are: race is not biological and natural but a social construct; US racism in the normal state and the US is normal and has been invented as a category intended to oppress people of colour and thus the laws and its institutions are racist and a variety of inequalities between white and people of colour; because of interest convergence, any changes in the status of people of colour will likely be of benefit to dominant white people; minority groups will experience differential racialisation through being associated with different sets of negative stereotypes; individuals can only be appropriately identified through their membership of multiple groups; people of colour are in the unique position to be able to speak on behalf of other people of colour regarding racism and its effects (also see critical approaches to research).

Data

It refers to the Information that is collected through research procedures (such as questionnaires, interviews, observations, etc.). Data can be qualitative or quantitative.

Data analysis

After data has been collected, it is analysed in order to attempt to answer the questions posed and addressed by the research. Theses analyses involve the processes by which data is organised so as to provide insights into regularities and differences in the behaviours of those people who are participating in the research. Data analysis can refer to many specific forms of analyses, for example, thematic analysis, content analysis and statistical analysis.

Data collection

Data collection is the gathering and recording of information through research methods such as interviews, observations and projective techniques.

Declarative mapping approach

The declarative mapping approach is an orientation within humanities and social sciences research. The approach has at its core the declarative mapping sentence (see below) which offers a clear description of the componential nature of a complex research domain (Gordley-Smith and Hackett, 2022). Research that is conducted within the declarative mapping approach is concerned with investigating a research domain by identifying the sub-components of the domain and using the declarative mapping sentence to investigate the interplay between these.

Declarative mapping sentence (DMS)

A declarative mapping sentence is an adapted form of a traditional mapping sentence (see *Mapping sentence [in a general sense]* later in this glossary). It has been developed by Paul Hackett (Hackett, 2014, 2016a, 2016b, 2016c, 2018) and is similar to its traditional counterpart except that it often does not include a range facet (it is traditionally the case that research designed within the facet theory rubric has a range facet that is specified prior to data collection). The declarative mapping sentence is usually used in qualitative and philosophical research.

Element

See *Facet element.*

Emic

The term emic relates to research that involves the "analysis of cultural phenomena from the perspective of one who participates in the culture being studied" (Merriam-Webster, 2022b).

Etic

Etic is a term that refers to the manner in which research is conducted where the research relates to involves "analysis of cultural phenomena from the perspective of one who does not participate in the culture being studied" (Merriam-Webster, 2022a). This contrasts to emic approaches.

Experiment

An experiment is a specific type of research study that is characterised by the manner in which it is designed. For a research study to be an experiment, it must be designed so as it has at least one independent variable and at least one dependent variable. An independent variable is a variable that is manipulated by the researcher in the experiment, and a dependent variable is a variable that changes its condition as a result of the manipulation of the independent variable. In order to be a true experiment, the study's design must also allow for the removal or control of extraneous variables or influences that may have an effect upon the state of the dependent variable (these may be other variables that are related to the independent variable or situational or background characteristics of the experimental setting or the person conducting the experiment). A true experiment must also have a control group which is made up of members that are similar to the sample group who are completing experiment, but instead of being presented with the independent variable, they receive some benign intervention that is similar to the independent variable. Experiments are nearly always associated with quantitative research. Experiments contrast with other forms of data gathering such as surveys, case studies and interviews, by being more rigorously controlled procedures.

Facet (a)

A facet can be thought of as a variable, construct or some other form of discretely identifiable component of a definition of a domain of research or important factor influencing a content that is under investigation. The facets that are specified for a research study, when taken together, constitute all of the variables that are of interest in the study. Facets are specified that are as much as possible mutually exclusive in terms of their combined influence within the research area.

Facet element

A facet element (or often just element) is a subdivision of a facet that is mutually exclusive with other elements of a facet, as near as possible to being mutually exclusive. Facet elements are not the means by which a mapping sentence is assessed but they constitute the structure of the phenomenon under scrutiny. Elements should not be confused with a facet range (however, a range facet will also possess elements).

Facet theory

Facet theory is a meta-theoretical approach to research design, implementation and analysis (Brown, 1985; Canter, 1985; Guttman, 1944; Hackett, 1983, 1995, 2014; Hildebrandt, 1986). Facet theory allows the mapping of some facet of an individual or group of people in reference to a process within a specified context. It achieves this through using a mapping sentence (facet theory's main research tool) the approach brings together a prescribed analysis of a research domain's contents with analytic approaches such as Smallest Space Analysis (SSA) (Lingoes and Borg, 1977) and Partial Order Scalogram Analysis (POSA) (Raveh and Landau, 1993): these analyses are multidimensional (Amar and Levy, 2014; Borg and Lingoes, 1987; Borg and Mohler, 2011). Facet theory integrates the design of research content with data analysis, which allows for a meaningful appraisal of a research domain and permits the possibility of theory development (see Brown, 2010).

Facet theory (qualitative)

I have developed the qualitative application of the facet theory approach to the study of human and non-human animal behaviour and experience over the past decade (Hackett, 2014, 2016a, 2016b, 2016c, 2018; St. Clair and Hackett, 2012). In this adaptation of the traditional facet theory orientation to research, I have moved away from the design of research and analysis of research findings that are numerical. Qualitative facet theory is therefore a structural template that facilitates the design and development of clearly and thoroughly specified research content. Qualitative facet theory then offers a framework for the analysis of narrative and other forms of thematic content analysis. Qualitative facet theory employs the declarative mapping sentence to allow the above claims to be realised.

Factors (factor analysis)

Martin and Bateson (2007) describe factor analysis as a statistical analysis method for uncovering groupings in data, where the main form of factor analysis is principal components analysis. The same authors state that factor analysis, "aim(s) to reduce the complex interrelationships between a large number of variables down to a smaller number of underlying factors that account for a large proportion of variance and covariance of the original variables" (Martin and Bateson, 2007, p. 116). Factors are components of the results that emerge from factor analysis (see Rencher and Christensen, 2012), which is a parametric graphical statistical analysis technique. The principal components analysis technique identifies a successive series of components, each of which explains a progressively smaller amount of the total amount of variation in the data set. There are two major forms of factor analysis: confirmatory (Brown, 2015) and exploratory (Fabrigar and Wegener, 2011). In the former, the approach is used to investigate a pre-existing or hypothesised factor structure that exists in the literature, whilst in the latter, there is no expected structure and the researcher explores possible structures based upon the data being analysed in the current study. Factor analysis and principal component analysis are similar to Smallestspace analysis. However, Factor analysis and principal components analysis are parametric statistical analysis techniques with factors that are produced through linear regression. Smallestspace analysis is an ordinal technique that yields non-linear facets which are the product of analysis of the inter-item similarities between the rankings of items.

Louis Guttman

Louis Guttman (1916–1987) was an American mathematician, sociologist and professor of social and psychological assessment. He is best known for his scholarship in the area of social statistics, for psychometric work (especially multidimensional scaling and his philosophical work on multidimensional analysis approaches), for the origination of the Guttman Scale and for the development of facet theory.

Hermeneutics

Hermeneutics is a branch of knowledge that is concerned with the content of narratives or texts. Specifically, hermeneutics is concerned with the interpretation of such texts and most commonly in the context of theological or literary texts.

Insight

Insight is a form of behaviour and learning found in animals. Animal insight grew out of the work of Wolfgang Köhler (1925) (see also, Mackintosh, 2015, for a review of this research; Chance, 1960). Insight learning studies are in opposition (or complementary to) associative learning and stimulus-response links (Mackintosh, 2015). Insight is indicated in an action that is based on a novel appraisal of a situation that results in behaviour that solves a problem. Insight has been investigated in bird behaviour, for example in pigeons, in a replication of Köhler's studies with chimpanzees (Epstein et al., 1984) and in other bird species (Emery and Clayton, 2004). Insight, in a more general sense, is the capacity to attain a deep and accurate understanding of an event, a phenomenon and so forth.

Information

A quality or quantity that resolves uncertainty (see *Data*).

Intelligence

There are different definitions of intelligence, but most involve notions of an individual's ability to develop and employ knowledge and skills.

Learning

Learning is demonstrated as a relatively permanent change in behaviour that is brought about due to experience. Learning is a major component or type of cognition and is often assessed in research into the cognitive abilities of individuals.

Linguistic

Linguistics is the scientific and systematically investigation of language through the use of a scientific method in the analysis of the form and structure of language, the meaning of language and language in the situation of its usage. Linguistics is also concerned with grammar, meaning and the sounds of a language.

Mapping sentence (in a general sense)

A mapping sentence is a formal statement of the research project that is being conducted in the format of a natural language sentence (see Hackett, 2014, 2018). A mapping sentence has three types of facets (more on these later), each linked with other using connecting words to form a sentence with a structure that approximates normal prose. The sentence suggests the expected interrelationship between the content facets within the context of the specified research inquiry. A mapping sentence is made up of three types or categories of information (these categories are called facets). The three types of facets are the background facet, the content facet and the range facet. Background facets specify detail of the events, people or objects to be classified or investigated in the research project. Background facets may also be sub-divisions of the population under investigation that you believe to be important in understanding the content of the domain under enquiry. Content facets specify the research domain that will be investigated in the project. Content facets are the major subdivisions of research content. For example: if your research project is investigating user's experience of a certain place, the facets will be the major aspects of place experience that have been identified in prior research to be influential in affecting users when they are in that situation. The more complex the investigation is the more content facets a project will contain. The range facet[1] specifies the overall orientation of the research project or is the measurement that will be taken in the research (e.g. in assessing user's experience of a place, the range facet may well be one of degrees of satisfaction with the various aspects of place as specified in the content facet overall customer satisfaction).

At this point, it will probably be useful to readers' understanding if we provide an example of a mapping sentence (Figure G1).

Mapping sentence mereology

A mapping sentence is a sentence written in ordinary English prose which contains facets and elements and where the

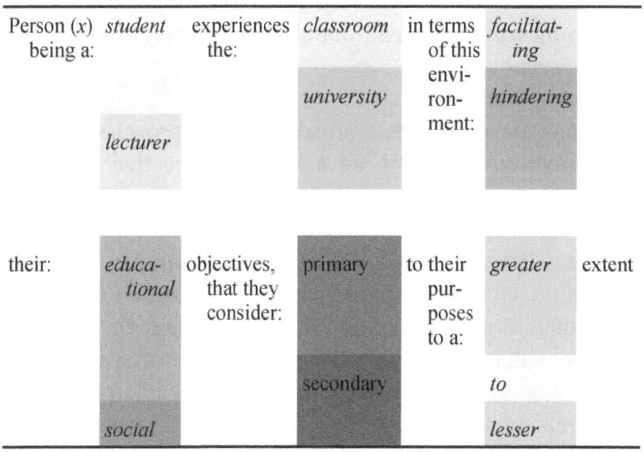

Figure G1 Traditional Mapping Sentence for Experience at University

facets are the major components of a research domain and the elements are the exhaustive, mutually exclusive conditions of the facets. "Mapping sentence mereology" is a term used in qualitative or philosophical facet theory and with the declarative mapping sentence. Mapping sentence mereology refers to the inter-connections and inter-relationships between the facets and the facet elements in a mapping sentence (usually declarative) including the connective components and functors (Varzi, 2016).

Mapping sentence ontology

A mapping sentence is a sentence written in ordinary English prose which contains facets and elements and where the facets are the major components of a research domain and the elements are the exhaustive, mutually exclusive conditions of the facets. "Mapping sentence mereology" is a term used in qualitative or philosophical facet theory and with the declarative mapping sentence. Mapping sentence ontology refers to the relationships and properties between the concepts and categories of a mapping sentence (usually declarative) including the connective components and functors.

Mapping sentence – declarative (see *Declarative mapping sentence*)

Mapping sentence – traditional (see *Mapping sentence [in a general sense]*)

Mapping techniques

Mapping techniques are a broad class of approaches to qualitative research that are a type of projective technique. They are also related to the concepts of cognitive mapping (e.g. see: Mira and Deus, 2005) and mental maps (e.g. Buzan, 2006). All mapping techniques or procedures involve the participant in the research to produce some form of a map. In this context, the term "map" can be understood as involving some form of spatial representation of some place, thing, state of affairs or entity. Depending on the focus of the research study, mapping techniques may or may not involve a respondent mapping an actual geographic location. In these approaches to data gathering, a respondent will be asked to draw a map of some kind. The map itself may take many forms but will usually be produced upon a large sheet of paper on which the participant writes words or phrases in a spatially significant way and then connects these with lines to demonstrate the associations between the terms.

Mixed method research

Mixed method research, as its name suggests, is a methodology that employs a mixture of both quantitative and qualitative methods in its design.

Multivariate behaviour research

Multivariate behaviour research is a form of enquiry that takes a variety of variables into account when it attempts to describe or infer the behaviour that is being investigated. Multivariate behaviour research does not address the effects of isolated independent variables upon equally singular dependent variables. Rather, research is designed and conducted to assess the effects of multiple variables upon one or more other variables (see Hahs-Vaughan, 2016; Coxon, 1982).

Online research

Online research is research that is conducted online. However, this simple description misses much of the complexity that exists in this type of research. Online social science research may be divided into two broad classes or types of research: research that is conducted into material that exists online and material that exists outside of the digital arena but is investigated using online technology. For instance, the former type of research may use information that exists on social media and investigate this and the latter may conduct a survey mounted and responded to online. It is obvious that projective techniques take the latter format. Many of the usual projective techniques are now available to use online, such as: writing obituaries for a specified thing or an event; a wide variety of mapping techniques can be created and manipulated in a digital medium, personification objectification and so forth (Accelerant Research, 2022).

Participant(s)

A participant (or participants) is the person (or persons) who responds, provides information or in some way takes part in a piece of research and whose information is analysed.

Personal construct theory

Personal construct theory is a personality and cognitive processing theory developed by George Kelly (REF). Personal construct theory (also personal construct psychology) is rooted in the idea that an individual understands their world through a series of evaluative dimensions that they develop through experience and which they carry around with them and apply in their daily experiences and encounters. These dimensions are termed "constructs" and are thought of, and represented by, a straight line that runs from one extreme of the construct to the opposite or contrasting extreme of the construct. Constructs are employed by an individual in all aspects of their lives and as they are developed through life are highly personal and result in individualised understanding of things, events, situations, other people and so forth (also see *Repertory grid technique*).

Personality

How individuals differ is one area of research and scholarship in psychology. Within this sub-domain, personality is one of the branches of study. Personality is therefore a psychological phenomenon that is a form of how individuals differ in terms of the typical ways in which they feel, think and act. Personality may be thought of a relatively enduring characteristic of an individual across different situations. Research into personality tends to focus upon two aspects of personality. The first of the main sub-areas of personality is concerned with how a person's personality characteristics come together as a functional whole. The second attends to the study of the individual characteristics of personality such as extraversion and extroversion. Several multi-componential models of personality have been developed.

Personality measures

Personality measures is a sub-area of psychometrics, which itself is concerned with the quantitative or statistical assessment of psychological processes, such as intelligence and various forms of behaviour. As personality is usually conceived as being made up of multiple relatively independent interacting components, personality tests and measures reflect this and often comprise multiple scores each of which reflect a specific aspect of personality. As with many other psychometric tests, personality tests are most often used in industrial and occupational psychology and also in clinical and educational settings.

Perspective taking

Perspective taking involves behaving in a way that can only be understood by assuming that the individual committing the behaviour is acknowledging/is aware of, a viewpoint other than just their own. However, the distinction between perspective taking and insight may be unclear.

Phrase

In language, a phrase is a small collection of words that form a conceptual unit with an idiomatic meaning and which usually are part of a clause.

Physical manipulation

This refers to the movement of objects, in order to achieve a goal or to complete a test task (a goal in the construction itself of using a construction for instrumental reasons to attain a higher-level goal).

Physical construction

This refers to building, forming of objects to help achieve a goal (a goal in the construction itself of using a construction for instrumental reasons to attain a higher level goal). The psychometric question may be asked as to the sufficiency of difference between physical manipulate tasks and physical construct tasks.

Placescape

A placescape is defined as "a representation of land or sea or city that emphasizes the locus and felt quality of place (above all its shape and scope)" (Casey, 2002, p. 352).

Processing speed

Is the time it takes for an individual to become aware, perceive and choose an action. Processing speed involves/includes the time taken by an individual to register a stimulus and to process this by referencing this to their extant schemata and attributing some sort of meaning to that and deciding upon committing an executive action.

Projective technique(s)

In social science and other forms of research with human subjects, the aim of an enquiry is often to increase understanding of a person's, or a group of people's thoughts and feelings towards an event, a state of affairs, concepts, social issues and some other topics. Researchers often use direct approaches to attempt to provide such understanding, such as questionnaires and so forth. Projective techniques are indirect methods that are typically used in qualitative or mixed-methods research. Projective techniques are indirect techniques that aim to facilitate or allow researcher to uncover and reveal a person's deep thoughts, motivations, feelings, beliefs, attitudes and values.

Arguably, it is valuable to attempt to understand these often effectual and irrational underlying psychological constructs (constructs which we may not be consciously aware of and therefore not amenable to direct questioning) as it is frequently these that impel our more obvious behaviours.

Because an individual may not have a clear understanding of their motives and subconscious influences upon these, projective techniques can be useful because they are tools that help to reveal the subconscious can be emotional and irrational in nature. Another advantage to be found in using projective techniques comes about because a person may experience difficulty in putting into words their motivations and desires, even when they are aware of these. Asking a direct question may produce useful and valid information, but it is also prone to respondents feigning good and offering socially or personally desirable answers to such enquiries. A person's deeper values and beliefs may be especially revealing of the person being questioned and there may be perceptual guards that prohibit such revelation. By actively involving a person in the physical construction of their responses, projective techniques assist the researcher to break through this resistance. Furthermore, projective techniques can be employed alongside other qualitative research approaches, such as direct questioning, and are very frequently used within focus groups.

Psychoanalytic theory (psychoanalysis)

Sigmund Freud was the originator of the theory of personality known as Psychoanalytic theory (REF). On the understandings present in psychoanalytic theory, personality is seen to dynamically consist of three components, the id, the ego and the super ego. The id is only present in the subconscious of an individual., whilst the ego and superego have conscious, pre-conscious and unconscious components. According to Freud, the conscious is the location of our thoughts and feelings, the pre-conscious is also called the subconscious and is our memories and representations of our past that we may recall, whilst the unconscious is the deepest level of the mind and comprises such aspects as primitive and instinctual desires and other drivers of our behaviour of which we are unaware (McCleod, 2014). The id is seen to comprise our survival and destructive instincts, the ego develops in childhood and keeps a check on the id whilst the superego contains morality and social responsibility. We will not go into more details as these are beyond the scope of this book. However, psychoanalytic theory has been developed as a therapeutic approach, and of particular interest to this book, it has spawned the development of a series of information-gathering approaches that, it is believed, are able to tap into the subconscious and even perhaps the unconscious parts of the mind.

Qualitative research

Qualitative research is also known as non-numerical research. The terms "qualitative" and "non-numerical" refer to the type of information (see definition earlier) that a research procedure yields. This type of research typically produces text, observational or narrative data and cannot be analysed statistically. Qualitative research typically attempts to produce rich understanding of the research context with findings that cannot be applied with confidence outside of the sample being investigated.

Quantitative research

Quantitative research is also known as numerical or statistical research. The terms "quantitative", "numerical" and "statistical" refer to the type of information (see definition earlier) that a research procedure yields. Quantitative research produces numerical data that may be analysed statistically. This form of research may be generalised outside of the sample used in the research with a specific level of confidence regarding the robustness of such generalisations. Quantitative data and research are typically not used to reveal respondents' rich personal meanings.

Range facet

The range facet is found in traditional quantitative facet theory (Borg, 1977). It specifies the range over which responses will be gathered: It explicitly states the values that responses will take. In all facet theory research, the range facet focuses on the research that is being conducted by specifying which information will be collected in order to test the veracity of the mapping sentences structure (Hackett, 2018).

Reliability

Reliability has a specific meaning in the context of research. On this meaning, reliability means the consistency and trustworthiness of a measure or assessment procedure. Therefore, what is meant by reliability is the extent to which a measure will produce the same result when it is applied on multiple occasions or at different times to the same sample.

Repertory grid

The repertory grid technique is a procedure for conducting social science research. The repertory grid approach gathers information from respondents in an interview-type setting in a grid format. The repertory grid technique was developed by George Kelly alongside personal construct psychology which he also originated. The basic notion behind personal construct psychology is that individuals understand the world around them through a series of dimensions that they develop through experience. These dimensions, or constructs, are linear and run one extreme pole to its opposite or contrasting pole. A person uses their individualised constructs to interpret things, events, situations, other people and so on. The repertory grid is a research procedure that enables a person to express how they see and understand the world by revealing their personal constructs whilst minimising interviewers induced biases. The repertory grid technique asks a person completing the construction of a grid about the similarities and differences between events often through the use of triadic sort procedures. The initial stages proceeds through a series of typical stages. In the first stage, the researcher chooses a series of what are called "elements" that are deemed to be examples of the topic being investigated. For example, the elements that would be included when investigating preferences for specific cars would be a variety of different cars thought to be representative of the choices available to the respondent. The second stage is the elicitation by the researcher of the personal constructs that relate to the differentiation of elements. In this stage, an attempt is made to understand how the respondent understands the content area of interest by asking the participant to identify the differences and similarities between elements, by using, for example, a triadic sort method. In this way, the respondent identifies how elements are different and similar to each other and reports these to the researcher. The procedure is repeated on different permutations of elements until no new constructs emerge. In the third stage, the use of a grid is employed. A grid is drawn with the names of the elements written above each column and with the rows labelled with, at the one end of the row, the one pole of a construct, and at the other end of the row, the contrasting pole's name. Respondents then rate each element in terms of writing a value from a scale of say 1 to 7, under each element column so that each row represents the extent of each construct for each element. In this procedure, the value of 1 would mean the element is very much represented by the left-hand pole of the construct, whilst a value of say 7, would mean the element is very much typified by the construct name at the right-hand of the row. Multivariate statistics are usually employed to analyse a grid's content (also see *Personal construct theory*).

Research

In the context of this book and within academia and professional settings (we are speaking primarily within the social science, business and humanities contexts), research is the systematic investigation into and the study of sources and events in the attempt to establish verifiable information in which confidence may be held. We are using the term in a more exacting sense than it is often used in everyday language to mean simply searching Google, or local shops. This is often un-systematised and non-stringent. Research is also more than simply asking questions. Research is the process of asking questions and interrogating these systematically within a theoretical context in order to provide answers. Research is more than questioning as it is rigorous and the procedure is explicit and documented.

Research bias

It refers to the factors that influence and distort the results that come out of a research study. Bias in research may come from a wide variety of sources. For example: selection of an inappropriate sample; factors associated with the locational or setting of the research; characteristics of the interviewer; inconsistencies in research procedures between different administrations of the research procedure; inaccurate data keeping or measurement taking, and so on. In statistics, bias refers to systematic errors or deviations in results. These are not the only sources of bias in research but serve to illustrate the breadth of the sources of potential bias and the many ways in which this may distort the findings of a research project.

Research design

Research design is more than simply the process of designing research. Research design is a formal approach to designing research that takes into account, and explicitly states, the relationships between variables and how research questions and hypotheses relate to each other. It also includes all of the features of a specific piece of research, including those within the research procedure itself and those external to the research process that may have an impact on the research results. The term "research design" means the strategy that a researcher uses in order to investigate an area of interest. Research design therefore involves the bringing together of the different aspects of a research study in a manner that effectively addresses the questions that are being investigated. Logic is used to achieve the above aims through the construction of a research design template or plan for collecting data, analysing this and considering measurement and statistical issues (see Abdi et al., 2009; Mitchell and Jolley, 2012; Spector, 1981; Spickard, 2016).

Research domain

The term "research domain" is synonymous with the term "research area". Both terms signify the content area addressed by a specific piece of research.

Respondent(s)

See *Participant(s)*.

Sort techniques

Sorting procedures are valuable research techniques that seek to uncover how individuals evaluate and categorise a wide variety of stimulus items. Simply put, sorting (or sort) procedures involve the presentation of stimulus materials to a participant who is then asked to sort this material into an arrangement (or arrangements) that are meaningful to them. In this way, a sort technique attempts to use a visually based activity to reveal a participant's mental concepts or the psychological constructs that guide their conscious and subconscious behaviours.

Sorting techniques can use a variety of different stimuli which the participant is asked to sort. These are often presented as cards with words printed on them, photographs or other pictorial forms, actually objects and so on, and may be physically presented or virtually presented on a tablet or other device. These techniques are all for eliciting knowledge that does not require direct questions from the researcher, although follow-up questions are often employed which ask the respondent to say more about the sort they have just completed.

In the procedure, items are presented to a participant and they are asked to sort these in meaningful ways, such as sorting these into groups and ranking items to identify order relationships between them. The precise way in which the sort is conducted will depend on the type of information the researcher is attempting to reveal. Furthermore, the researcher may ask the participant to sort the material in a certain way (i.e. into a specified number of groups) or ask them to impose their own sorting protocols that reflect their understanding of the similarities and differences between the items.

Spatial ability

Spatial ability is evidenced in acting in a manner that suggests that space and volume are understood.

Stage sequential/stage sequential reasoning

Acting stage sequentially involves undertaking requisite tasks in the necessary order to enable the attainment or achievement of a goal or outcome. Stage sequential reasoning is demonstrated in an individual's behaviour when the person links a series of novel actions together in order to attain a goal.

Statistical analysis/research

Statistical research and analysis are approaches to inquiries that are not typically found used in conjunction with projective techniques. This is due to statistical approaches being concerned with numerical responses to direct questions, and projective technique research is qualitative in terms of the data collected and deep understanding is sought rather than number-based responses (Millsap and Maydue-Olivares, 2009). Statistical methods and statistical forms of analysis are of two broad classes: descriptive (Jones and Goldring, 2022) and inferential (Macinnes, 2022). The former of these uses numerical values to describe responses, events, characteristics of an individual and so on. These include frequency counts, averages, ranges of responses and other, and these are sometimes used in conjunction with projective approaches and the information they produce. The other major form of statistical analysis is inferential and as the name suggests, these statistics compare the numerical data collected and are used to infer the numerical significance of the similarities and differences between the responses from groups of respondents. Inferential statistics include t-tests, correlation tests, multi-dimensional procedures and so on (Borg et al., 2018; Sabo and Boone, 2013; Meyers et al., 2010).

Subject(s)

See *Participant*.

Test battery

In psychological assessment, individuals are usually tested using a variety of different, though related, tests. This is the case in the assessment of intelligence and other skills and abilities. Together, these tests are known as a test battery.

Theory

The word theory has slightly different meanings, but in this book, a theory is taken to be related to a set of ideas which are used to explain other events or things. Theories often are general in their nature and are independent to the events, states of affairs and so on, that they are being employed to explain. In science, theories are often used in the above sense in order to explain the natural world in a manner that avails itself to investigation and verification, refutation or amendment through the process of repeated investigation of the theory's ability to provide explanations of empirical observations.

Traditional mapping sentence

See *Mapping Sentence in a General Sense.*

Transference

Transference implicates the role of longer-term memory and learning. Transference involves taking an experience from one situation or task and applying it to another situation in which the behaviour would not have been observed and would not have been expected to elicit this behaviour.

Validity

Validity refers to the instance when something is rooted in truth, fact, supported by a law and so on. Perhaps of more direct relevance to the present writing, validity means to something that is supported by theoretical rationale or empirical evidence. Moreover, validity implicates adequacy and appropriateness in terms of any conclusion that is derived from research or another form of assessment (American Psychological Association, 2018).

Variables

Variables are components of experimental situations and a part of the language used to talk about and describe much quantitative research. Variables are events or other entities that may be controlled, altered or measured in an experiment. More specifically, in experiments, variables may take one of several types. Independent variables are the variables that are manipulated in the course of an experiment. Dependent variables are the outcome variables or the variables that alter due to the manipulation of the independent variables. Controlled variables are the variables that are held constant in an experiment in order to remove their effects upon the dependent variables. Extraneous variables are the variables that have an effect upon the dependent variables but are not taken into account when considering the effects of the independent variables. Extraneous variables are a source of error in an experiment.

Visual comprehension

Visual comprehension requires an individual to act in a way that demonstrates that they are able to perceive and act in a manner that demonstrates visual recognition and action based upon this visual recognition.

Visual spatial

In the context of this book, visual-spatial refers to a form of intelligence. So conceived, visual-spatial tests are usually concerned with the ability of an individual to mentally manipulate objects in three dimensions or to conceive two-dimensional representations in three dimensions.

Word

A word may take two forms, a written or a spoken form. In the former, it consists of one or more letters, and in the latter, it comprises one or more sounds that symbolise and communicate a specific meaning. A word is the smallest written or spoken component that can stand alone and express an applied, a realistic or an objective meaning. The parts of written words are graphemes, and the components of spoken words are phonemes. However, the meaning of words has been much debated (see Di Sciullo and Williams, 1987; Bromberger, 2011).

Working memory

Memory is conventionally divided into three components: working memory; short-term memory and long-term memory. Working memory takes the form of information that is held in the conscious mind that enables one to act. It is the most temporary form of memory. Traditionally, working memory has been seen as being limited in its capacity (usually to seven plus or minus items (Miller, 1956 REF), but more recently, working memory has been posited as being a restricted and controlled resource that is distributed across all of the items that are being kept in memory (see Ma et al., 2014).

References

Abdi, H., Edelman, B., Valentin, D., and Dowling, W.J. (2009). *Experimental Design & Analysis for Psychology*. Oxford: Oxford University Press.

Accelerant Research (2022). *Online Projective Exercises Breathing New Life Into Classic Qualitative Research Techniques*. The Greenbook Directory. Retrieved from www.greenbook.org/marketing-research/online-projective-exercises-37076

Amar, R., and Levy, S. (2014). SSA: Similarity Structure Analysis. In A.C. Michalos (ed.), *Encyclopedia of Quality of Life and Well-Being Research* (pp. 6306–6313). New York: Springer.

American Psychological Association (2018). *APA Dictionary of Psychology*. Retrieved from https://dictionary.apa.org/cognition

American Psychological Association (2022). *APA Dictionary of Psychology*. Retrieved from https://dictionary.apa.org/cognition

Borg, I. (1977). Some Basic Concepts in Facet Theory. In J.C. Lingoes, E. Roskam, and I. Borg (eds.), *Geometric Representation of Relational Data* (pp. 65–102). Ann Arbor: Mathesis.

Borg, I., Groenen, P.J.F., and Mair, P. (2018). *Applied Multidimensional Scaling and Unfolding (Springer Briefs in Statistics)*. New York: Springer.

Borg, I., and Lingoes, J. (1987). *Multidimensional Similarity Structure Analysis*. New York: Springer.

Borg, I., and Mohler, P.P. (2011). *Trends and Perspectives in Empirical Social Research*. Berlin: Walter de Gruyter.

Bromberger, S. (2011). What Are Words? Comments on Kaplan (1990), on Hawthorne and Lepore, and on the Issue. *Journal of Philosophy*, 108, 485–503.

Brown, J. (1985). An Introduction to the Uses of Facet Theory. In D. Canter (ed.), *Facet Theory: Approaches to Social Research*. New York: Springer Verlag.

Brown, J.M. (2010). Designing Research Using Facet Theory. In J.M. Brown and E.A. Campbell (eds.), *The Cambridge Handbook of Forensic Psychology* (pp. 795–802). Cambridge: Cambridge University Press.

Brown, T.A. (2015). *Confirmatory Factor Analysis for Applied Research* (2nd edition) (Methodology in the Social Sciences). New York: The Guilford Press.

Buzan, T. (2006). *The Ultimate Book of Mind Maps*. New York: Harper Collins Publishers.

Canter, D. (ed.) (1985). *Facet Theory: Approaches to Social Research*. New York: Springer Verlag.

Casey, E.S. (2002). *Representing Place: Landscape Painting and Maps*. Minneapolis: University of Minnesota Press.

Chance, M.R.A. (1960). Köhler's Chimpanzees: How Did They Perform? *Man*, 60, 30–135.

Coxon, A.P.M (1982). *The User's Guide to Multi-Dimensional Scaling With Special Reference to the MDS (X) Library of Computer Programs*. London: Heinemann Educational.

Di Sciullo, A.-M., and Williams, E. (1987). *On the Definition of Word*. Cambridge, MA: MIT Press.

Emery, N. J., and Clayton, N. S. (2004). The mentality of crows: Convergent evolution of intelligence in corvids and apes. *Science*, 306(5703), 1903–1907. Retrieved from https://search-proquest-com.ezp.lib.cam.ac.uk/docview/213609218?accountid=9851

Epstein, R., Kirshnit, C.E., Lanza, R.P., and Rubin, L.C. (1984). 'Insight' in the Pigeon: Antecedents and determinants of an Intelligent Performance. *Nature*, 308(5954), 61–62.

Epstein, R., Patai, E., Julian, J., and Spiers, H.J. (2017). The Cognitive Map in Humans: Spatial Navigation and beyond. *Nature Neuroscience*, 20, 1504–1513. https://doi.org/10.1038/nn.4656

Fabrigar, L.R., and Wegener, D.T. (2011). *Exploratory Factor Analysis (Understanding Statistics)* (1st edition). Oxford: Oxford University Press.

Gordley-Smith, A., and Hackett, P.M.W. (2022, January 18–21). The Declarative Mapping Sentence Method for Integrating Disparate Forms of Research Related to Biosocial Medical Anthropology and Covid-19, Mobilising Methods in Medical Anthropology: Biosocial Medical Anthropology and Covid-19. In *Re-thinking Concepts and Methods in Pandemic Times*, Royal Anthropological Institute, London, UK.

Guttman, L. (1944). A Basis for Scaling Quantitative Data. *American Sociological Review*, 9(2), 139–150.

Hackett, P.M.W. (1983). *Observations on Blink Rates in Ferruginous Duck (Aythya nyroca) in a Flock of Mainly Mallard (Anas Platyrhynchos)*. Working Paper.

Hackett, P.M.W. (1995). *Conservation and the Consumer: Understanding Environmental Concern*. London: Routledge.

Hackett, P.M.W. (2014). *Facet Theory and the Mapping Sentence: Evolving Philosophy, Use and Application*. Basingstoke: Palgrave Macmillan.

Hackett, P.M.W. (2016a). *Psychology and Philosophy of Abstract Art: Neuro-aesthetics, Perception and Comprehension*. Basingstoke: Palgrave Macmillan.

Hackett, P.M.W. (2016b). *The Perceptual Structure of Three-Dimensional Art*. Heidelberg: Springer.

Hackett, P.M.W. (2016c). Facet Theory and the Mapping Sentence As Hermeneutically Consistent Structured Meta-Ontology and Structured Meta-Mereology. *Frontiers in Psychology: Philosophical and Theoretical Psychology*, 7, 471. https://doi.org/10.3389/fpsyg.2016.00471

Hackett, P.M.W. (2018). Declarative Mapping Sentence Mereologies: Categories From Aristotle to Lowe. In P.M.W. Hackett (ed.), *Mereologies, Ontologies and Facets: The Categorial Structure of Reality*. Lanham, MD: Lexington Publishers.

Hackett, P.M.W. (2019). *The Complexity of Bird Behaviour: A Facet Theory Approach*. Cham, CH: Springer.

Hahs-Vaughan, D.L. (2016). *Applied Multivariate Statistical Concepts*. London: Routledge.

Hildebrandt, L. (1986). A Facet Theoretical Approach for Testing Measurement and Structural Theories: an Application of Confirmatory Mds. In R.J. Lutz (ed.), *NA - Advances in Consumer Research Volume 13* (pp. 523–528). Provo, UT: Association for Consumer Research.

Jones, J.S., and Goldring, J. (2022). *Exploratory and Descriptive Statistics (The SAGE Quantitative Research Kit)*. Thousand Oaks, CA: Sage Publishers.

Kohler, W. (1925) 1925. *The Mentality of Apes*. London: Kegan Paul, Trench, Trubner.

Lingoes, J.C., and Borg, I. (1977). Identifying Spatial Manifolds for Interpretation. In J.C. Lingoes (ed.) (1979), *Geometric Representations of Relational Data*. Ann Arbor: Mathesis Press.

Ma, W.J., Husain, M., and Bays, P.M. (2014). Changing Concepts of Working Memory. *Nature Neuroscience*, 17(3), 347–356.

MacInnes, J. (2022). *Statistical Inference and Probability (The SAGE Quantitative Research Kit)*. Thousand Oaks, CA: Sage Publishers.

Mackintosh, N.J. (2015). *Animal Learning*, Encyclopaedia Britannica, Encyclopaedia Britinnica Inc. https://www.britannica.com/science/animal-learning/Insight-and-reasoning

Martin, P., and Bateson, P. (2007). *Measuring Behaviour: An Introductory Guide*. Cambridge: Cambridge University Press.

McCleod, S. (2014). Psychodynamic Approach. *Simply Psychology*. Retrieved from www.simplypsychology.org/psychodynamic.html

Merriam-Webster (2022a). Emic *adjective*. Merriam-Webster Online Dictionary. Retrieved from www.merriam-webster.com/dictionary/emic

Merriam-Webster (2022b). Etic *adjective*. Merriam-Webster Online Dictionary. Retrieved from www.merriam-webster.com/dictionary/etic

Meyers, J.L., Well, A.D., and Lorch, R.F. (2010). *Research Design and Statistical Analysis* (3rd edition). London: Routledge.

Miller, G.A. (1956). The Magical Number Seven, Plus or Minus Two: Some Limits on Our Capacity for Processing Information. *Psychological Review*, 63, 81–97.

Millsap, R.E., and Maydue-Olivares, A. (eds.) (2009). *The SAGE Handbook of Quantitative Methods in Psychology (Sage Handbooks)*. Thousand Oaks, CA: Sage Publishers.

Mira, R.G., and Deus, J.E.R. (eds.) (2005). *Environmental Perception and Cognitive Maps: A Special Issue of the International Journal of Psychology, 40(1)*. New York: Psychology Press.

Mitchell, M.L., and Jolley, J.M. (2012). *Research Design Explained*. Belmont, CA: Wadsworth Publishing.

Raveh, A., and Landau, S.F. (1993). Partial Order Scalogram Analysis with Base Coordinates (POSAC): Its Application to Crime Patterns in All the States in the United States. *Journal of Quantitative Criminology*, 9(1), 83–99.

Rencher, A.C., and Christensen, W.F. (2012). *Methods of Multivariate Analysis* (3rd edition). New York: John Wiley and Sons, Inc.

Sabo, R., and Boone, E. (2013). *Statistical Research Methods: A Guide for Non-Statistician*. Cham, CH: Springer.

Spector, P.E. (1981). *Research Designs (Quantitative Applications in the Social Sciences)*. Thousand Oaks, CA: SAGE Publications, Inc

Spickard, J.V. (2016). *Research Basics: Design to Data Analysis in Six Steps*. Thousand Oaks, CA: SAGE Publications, Inc

St. Clair, K.L., and Hackett, P.M.W. (2012). Academic Challenge: Its Meaning for College Students and Faculty. *Journal on Centers for Teaching and Learning*, 4, 101–117.

Thomasson, A. (2018). Categories. In *The Stanford Encyclopedia of Philosophy* (Spring 2018 edition). Edward N. Zalta (ed.).

Retrieved from https://plato.stanford.edu/archives/spr2018/entries/categories/.

Tolman, E.C. (1948). Cognitive Maps in Rats and Men. *Psychological Review*, 55(4), 189–208. https://doi.org/10.1037/h0061626. PMID 18870876.

Varzi, A. (2016). *Mereology: The Stanford Encyclopedia of Philosophy* (Winter 2016 edition). Edward N. Zalta (ed.). Retrieved from https://plato.stanford.edu/archives/win2016/entries/mereology/

Index

For Product Safety Concerns and Information please contact our EU representative GPSR@taylorandfrancis.com Taylor & Francis Verlag GmbH, Kaufingerstraße 24, 80331 München, Germany